AMERICA'S SPACE FUTURES
DEFINING GOALS FOR SPACE EXPLORATION

Eric R. Sterner, editor

Table of Contents

Introduction: Competing Visions
Eric R. Sterner — 1

The Cislunar Solution: A 21st Century Space Strategy
James A. Vedda — 10

Advancing U.S. Geopolitical and International Interests in Space
Scott D. Pace — 32

Ensuring Space Leadership: A 21st Century R&D Investment Strategy
William B. Adkins — 61

Achieving Cheap Access to Space, the Foundation of Commercialization
Charles M. Miller — 91

Conclusion: The Vision Thing
Eric R. Sterner — 125

About the Authors — 142

About the George C. Marshall Institute — 145

Introduction: Competing Visions
Eric R. Sterner

Do we need another volume on space policy? It is not just a rhetorical question, but also likely one an editor of any such essay collection should answer, if for no other reason than to convince the reader to keep going. Clearly, our answer at the George C. Marshall Institute is "yes." But, why? After all, it is not difficult to find informed commentary about the space program, ranging from National Research Council (NRC) reports to snarky observations in blogs or at conferences -- and everything in between.

The National Aeronautics and Space Administration (NASA) is in a state of perpetual uncertainty. Since, at the moment, the success of the U.S. civil space program is largely defined by the agency's state of affairs, such irresolution bodes ill for the nation's prospects of achieving its most important goals, to the degree that it can settle upon them. Moreover, constant churn around the agency undermines the cost-effectiveness of its activities, which require predictability but cannot rely on consistent budgets or guidance.

Criticism that NASA lacks a vision or guiding principle around which to organize its priorities, plans, and programs has become commonplace. Following the loss of the space shuttle Challenger in 1986, the

Reagan Administration created the National Commission on Space "to formulate a bold agenda to carry America's civilian space enterprise into the 21st century."[1] The Commission laid out a vision for making the solar system—and not just Earth—humanity's home in the cosmos and identified several steps for achieving it. It envisioned completing and operating a permanent space station, followed by new space transportation vehicles that would radically lower costs, modular transfer vehicles between Low Earth Orbit (LEO) and other destinations in the solar system, LEO spaceports, a lunar outpost, shipment of shielding mass from the moon, a spaceport in lunar orbit, initial operation of an Earth-Mars transportation system for robotic precursors, flight of an Earth-Mars cycling spaceship, human exploration of planetary moons, and development of Martian resources. Perhaps due to its ambitiousness in a time of declining budgets, that vision failed to satisfy critics of the nation's space program. Arguably, the federal government has tried to follow this plan by completing the space station and attempting—but failing—to replace the space shuttle with a more affordable alternative. Still, it remains largely stuck.

In 1990, President George H.W. Bush, concerned by continuing complaints about a perceived lack of direction in the space program, convened an *Advisory Committee on the Future of the U.S. Space Program*. The committee identified nine issues with the program, the first of which was "the lack of national consensus as to what should be the goals of the civil space program and how they should in fact be accomplished."[2] The committee also concluded "NASA is currently over committed in terms of program obligations relative to resources available."[3]

---

1 National Commission on Space, *Pioneering the Space Frontier* (New York: Bantam Books, May, 1986), http://history.nasa.gov/painerep/begin.html.
2 *Report of the Advisory Committee on the Future of the U.S. Space Program* (Washington, DC: Government Printing Office, December, 1990): 2.
3 *Report of the Advisory Committee on the Future of the U.S. Space Program*, 2.

Fast-forward nearly a quarter century past the loss of the space shuttle Columbia. The accident investigation board specifically singled out the Advisory Committee's report and strongly noted, "The U.S. civilian space program has moved forward for more than 30 years without a guiding vision, and none seems imminent."[4] In other words, on that score, little had changed.

Frustrated with the chronic nature of the problem, in 2011, Congress asked for an independent assessment of NASA's strategic direction. In response, the National Research Council produced a report, *NASA's Strategic Direction and the Need for a National Consensus* at the end of 2012. Its observations and conclusions are achingly familiar: "[T]here is no strong, compelling national vision for the human spaceflight program, which is arguably the centerpiece of NASA's spectrum of mission areas. The lack of national consensus on NASA's most publicly visible mission, along with out-year budget uncertainty, has resulted in the lack of strategic focus necessary for national agencies operating in today's budgetary reality. As a result, NASA's distribution of resources may be out of sync with what it can achieve relative to what is has been asked to do."[5]

Not surprisingly, the committee further identified a mismatch amongst NASA's budget, portfolio of missions, facilities, and staff. These implementation problems, along with those that virtually every large technology development program experience, conspired to exacerbate debates about any vision that a President had offered. According to the NRC, part of that continuing problem flows from the lack of clear and consistent strategic guidance from policymakers. In its absence, NASA has attempted, and largely failed, to chart a strategic course for the nation's

---

[4] Columbia Accident Investigation Board, *Report, Volume 1* (Washington, DC: Columbia Accident Investigation Board, August, 2003):210.
[5] Committee on NASA's Strategic Direction, National Research Council, *NASA's Strategic Direction and the Need for a National Consensus* (Washington, DC: The National Academies Press, 2012): 1.

space program. But, as the NRC report notes, "Absent such a consensus, NASA cannot reasonably be expected to develop enduring strategic priorities for the purpose of resource allocation and planning."[6]

A private organization, The Space Foundation, came to similar conclusions at roughly the same time. Its report, *Pioneering: Sustaining U.S. Leadership in Space*, observes, "As the space program has evolved, we have witnessed frequent redirection and constantly shifting priorities at NASA, mixed signals from Congress and the administration, organizational conflicts, and the lack of a singular purpose, resulting in a space agency without a clear, stable direction … NASA needs to embrace a singular, unambiguous purpose that leverages its core strengths and provides a clear direction for prioritizing tasks and assigning resources."[7]

Quite simply, despite decades of stunning accomplishments, ranging from deep outer space probes and scientific missions, to building and deploying one of the most complex and difficult engineering feats in human history (the International Space Station), the agency, and hence the U.S. civil space program, continues to suffer from chronic problems that have afflicted it since the end of the Apollo program. Therein lies the root of the problem. The United States has not been able to establish and sustain a consensus around a unifying purpose for the civil space program since the Kennedy Administration's commitment to beat the Soviets to the moon. Even that national goal was challenged, often by members of Congress and occasionally by President Kennedy himself, suggesting that it may not has been as "unifying" as we remember.

Johnson deferred major post-Apollo decisions to Nixon, whose administration did not embrace civil space as a national priority. Budget

---

[6] National Research Council, *NASA's Strategic Direction*, 39.
[7] Space Foundation, *Pioneering: Sustaining U.S. Leadership in Space* (Colorado Springs, CO: Space Foundation, 2012): 1.

cuts that began under Johnson continued, despite the success of the Apollo program, while the Nixon and Ford administrations sought to introduce a more rational program planning process and pursue a "balance" of manned and unmanned programs. Nixon decided to proceed with the space shuttle program for several reasons, casting it both as a long-term money-saver and contributor to détente. President Carter was not an advocate of civil space and his Vice President, Walter Mondale, had been one of its leading Congressional opponents. As a result, NASA remained largely on the same autopilot that a disinterested Nixon Administration had put it on earlier in the decade.

The Reagan Administration took a greater interest in civil space, seeing geopolitical and economic value in space activities. Following the space shuttle's early missions, the President initiated the space station Freedom program (1984), the National Aerospace Plane (1986), and recommitted the country to an ambitious future in space following the loss of the space shuttle Challenger in 1986. President George H.W. Bush, who had been Reagan's Vice President, offered an Apollo-like, mission-focused concept for the future of the space program in 1989, on the 20th anniversary of the first moon landing. The Space Exploration Initiative (SEI) would return Americans to the moon and venture on to Mars. While Congress did not reject the vision outright, it declined to fund the initiative at a time of heightened concern about growing budget deficits.

The Clinton Administration, which came into office in 1993, eventually cancelled the SEI and the National Aerospace Plane, while threatening to cancel the International Space Station before reorganizing it as a foreign policy initiative that would strengthen U.S.-Russian relations, advance certain nonproliferation goals, and demonstrate Russia's emergence from Communist authoritarianism.

Following the 2003 loss of NASA's oldest shuttle orbiter, Columbia, the George W. Bush Administration offered the Vision for Space Exploration, or VSE. Like the SEI, the VSE called for a return to the moon before sending human to Mars. Unlike SEI, which was not initially constrained by serious budget projections, the VSE was supposed to fit within a period of modest growth in NASA's budget. The administration announced that it would end the shuttle program and complete the International Space Station, which would free up funds for the VSE. Although Congress endorsed the VSE on a bipartisan basis, neither the administration or Congress provided enough resources to accommodate the cost of returning the space shuttle system to flight, rebuilding from natural disasters that affected NASA's infrastructure, or pursuing the multiple programmatic and scheduling goals associated with the VSE.

The Obama Administration, which arrived in 2009, cancelled the VSE and its flagship programs, proposing instead to develop technology, visit an asteroid, and expand government subsidies for companies seeking to stimulate the emergence of a commercial market in human spaceflight. Congress adjusted the administration's approach, resurrecting flagship VSE programs under a new name and declining to fund most of its technology programs, but affirming its plan to subsidize the private development of new low-Earth orbit space capabilities that could meet the need to sustain the International Space Station.

In short, since the Apollo program, few presidents have offered comparable unifying, mission-focused visions. Congress rejected one and policymakers (the legislative and executive branches combined) failed to provide sufficient resources for the other. On both occasions, each president's successor reversed the nation's course in space.

Given this brief history, it is reasonable to ask whether any recommendation to build consensus is truly meaningful. Agreement is always desired, but it may already exist after a fashion. Various interest groups seek to influence the United States' course in space. The competing and divergent priorities and interests of those groups have made it politically necessary to satisfy them all in order to conduct civil space activity. Predictably, that produces a program, which occasionally accomplishes great things, but remains less than the sum of its parts. In light of past failures to implant a multi-decadal, guiding vision in the space program's DNA, it appears there is, in fact, a political consensus for the space program: the status quo, as deficient as it may be in some eyes, is politically preferable to any unified, destination-focused course of action. The policymaking community and process are firmly resolved to remain indecisive on this score. Of course, that does not leave NASA with nothing to do; the post-Apollo decades witnessed a range of contributions to science, exploration, human spaceflight, and even American geopolitical leadership. No doubt, the agency will continue in that tradition. So, the question for policymakers is not whether NASA has a future, but, rather, how should they define the agency's role in America's future?

This dilemma prompted the Marshall Institute to produce this collection of essays. Most simply, we asked what a space program designed to achieve a single goal—rather than satisfy multiple divergent interests—should look like. We asked experts in their fields to contemplate alternative visions for a national space program based on different sets of national goals: spreading human civilization into the solar system; strengthening the U.S. position geopolitically; expanding space-related commercial activity; and, developing new technologies. Each author was charged with playing devil's advocate, designing an organization, program, and policies to pursue the assigned goal without regard to space history, politics, or other goals. In effect, we wanted to pursue each vision to its logical, yet unrealistic, extreme.

Of course, we recognize that even unified and focused goals are not mutually exclusive. Traveling to the moon, for example, required the development of new and incredibly advanced technologies as well as a new and larger cadre of trained scientists and engineers, not to mention an increase in our scientific understanding of cislunar space and the human body under abnormal stresses. Those technologies, people, and knowledge, of course, also benefitted the American economy, science, and geopolitical position independent of their contributions to the Apollo program. In the same vein, adding to the sum total of human knowledge requires new technologies, which could similarly find their way into the economy and other areas of human activity. That said, setting those goals—reaching the moon for geopolitical reasons or advancing scientific research—is not likely the most efficient way of achieving other goals, such as developing new technology for economic productivity or training new cadres of personnel with advanced training in science, technology, engineering, or math.

Taken collectively, the purpose of these essays is not to develop or identify a set of recommendations for the future of the American space program. Rather, it is to sharpen distinctions and identify the tradeoffs and commonalities that any space program seeking to pursue multiple goals must make. This should be a prerequisite for acting on the recommendations of the aforementioned studies, commissions, and reports that the United States develop a unifying goal. Surely, settling on a national vision — if such proves possible—should consider the opportunity costs of foregoing the alternative futures.

The authors are engaged in thought experiments, not in advocacy. Their essays may, or may not, represent their personal conclusions or recommendations for the U.S. space program *today*. We specifically asked the authors to think "out of the box" and begin with a clean slate, rather than be bound by the hardware, technology, markets, or political realities that define the current policymaking environment. The Institute hopes that this collection will enlighten policymaking by helping the policy community confront, and question, assumptions. It highlights the pros and cons of pursuing multiple national goals without setting priorities among them and offers alternatives for policymakers to consider as they wrestle with the challenge of building a future for the United States in space.

The Cislunar Solution: A 21st Century Space Strategy

James A. Vedda

For much of the time since the end of the Apollo era, participants in the space community and members of the interested public have been frustrated by the nation's inability to choose a coherent path for space exploration and development, make a commitment to its implementation over the long term, and fund it appropriately. Despite the aggravation caused by this indecision and instability, more change is in order.

Why propose such a thing? The short answer is that this is a high-risk, high-cost, long-term endeavor, so it is more important to get it right than to do it fast. The long answer is that we need to redefine what it means to be a spacefaring nation in the 21st century. (Hint: Space is more than just a prestige activity and another path for relaying electromagnetic signals.)

The end of the Cold War at the beginning of the 1990s should have been a wake-up call that it was time to rethink the justification and strategy for human spaceflight. But most people missed the call. Some of those who did recognize it perceived it as a death knell, perhaps believing that U.S. human spaceflight was merely surviving on the momentum of the space shuttle and space station programs.

What changed for human spaceflight at the end of the Cold War? Much has been said and written on this topic, so let me boil it down. During the Cold War, human spaceflight fulfilled two major national purposes: 1) it contributed to prestige, and 2) it was a research and development project that stimulated technology development and the industrial base. In a few specific locations around the country, it was also a jobs program. At the time, this was sufficient to justify the investment and the risk. After the Soviet threat disappeared, these purposes still existed, but they were *no longer sufficient* to justify indefinite continuation of the program in the United States – although they may still be sufficient for emerging space-faring nations such as China and India.

That raises the question of what comes next, which has been and continues to be asked all around the space community and among policy-makers in Washington. Answering this question, as it pertains to the U.S. government's role, confronts pitfalls driven by short-term thinking coupled with partisanship and parochialism. The discussion quickly devolves into a shopping list of solar system destinations that we think astronauts will be able to visit within the next decade or two. We then pretend that the time, resources, and risk involved in the effort can be justified by the tech spinoffs, the aerospace jobs created, and the inspiration it will give to our youth. The goal of sustainable space activities that have purpose and generate value (beyond scientific findings) gets lost, as does the realization that creating a diverse and thriving in-space infrastructure will be a multi-generational and multi-national effort.

Creating a productive future in space exploration and development requires difficult technical, economic, and political tradeoffs. We want to advance our technologies and capabilities, but people disagree which approach offers the best chance of success: tying technical developments to specific missions, or more general investment in improving the state of the art in particular areas. A specific mission presents a clear goal, well-defined requirements, and often a firm schedule, but it is inflexible, doesn't leave much room for general experimentation, and usually lacks a plan for establishing a long-term infrastructure. The final product may be a system with very narrow applicability that quickly gets discarded even if it is highly successful (a prime example being the Saturn 5 launch vehicle). On the other hand, too much reliance on undirected research efforts could be deficient in usable results as the quest continues for even better, more advanced, more elegant outcomes. Striking the right balance between the two approaches is a subjective exercise, more art than science.

Ideally, we would like to establish a series of significant milestones and achieve them as quickly as possible. This would keep the workforce busy, attract fresh talent to the effort, and demonstrate accomplishments that help sustain public and political support. But pushing programs onto a fast track without good reason and adequate funding could be their undoing. Exaggerated promises and unmet milestones will have negative repercussions. Even if there is short-term gain, a hastily executed program could undermine long-term utility. Strategic planners struggle to balance these many considerations.

Moon, Mars, Asteroids… What's Next?

A human spaceflight strategy should not be devised in isolation from everything else. To do so would be like asking, "Assume a human spaceflight program. What should we do with it? Where should we send astronauts that will keep everyone interested?" This is a good way to make the endeavor look like a stunt or an athletic competition in which a crew reaches a "finish line," after which everyone's interest quickly wanes, crippling follow-on efforts.

To begin the process of setting sustainable long-term goals and crafting a strategy to carry them out, the first question that should be on the lips of policymakers is: "What program of space exploration and development would best serve U.S. national interests?" Then, *after* a reasonable answer to that question has been developed, the next question is: "What is the role of human spaceflight in this program?"

Before setting dates for human missions to asteroids or Mars, the U.S. needs to define the purpose of such missions as part of a larger strategic plan in the national interest. A plan of great scope and duration is extraordinarily difficult to formulate, gain approval for, and sustain. But enduring success in a resource-constrained environment demands that we undertake the difficult process of formulating and winning approval for a plan to expand human and robotic activity throughout the Earth-Moon system and then to other parts of the solar system. This cannot work if we try to substitute tools and tactics ("Let's build a new rocket that can fly beyond low Earth orbit") for goals and strategies ("Let's create new knowledge, new capabilities, and long-term benefits"). If we make this mistake, we risk expending too many of our resources on designing, building, and testing flight hardware with little left to operate it or develop payloads for it, thus defeating the purpose of the whole exercise.

In recent years, NASA's human spaceflight program has displayed some shift in its focus toward capabilities, knowledge, and experience, while reducing the dominance of destination-driven planning. The rationale for making a capabilities-driven approach central to the exploration and development of the solar system can be found throughout human history. What has motivated human societies, usually at great cost and risk, to undertake major migrations to, or activities in, unfamiliar and challenging environments? It comes down to two things. First, they go where the resources are. Humans in search of precious minerals, raw materials, and energy, and the wealth they bring, have explored the most hazardous environments on Earth, including the ocean floor, the polar regions, treacherous terrain, and underground mines. Valuable discoveries have spawned economic booms and determined human migration and settlement patterns. Second, they search for new avenues to solve their problems and improve their living conditions. There are many examples of communities of people moving to escape a deteriorating environment (famine, drought, overcrowding, etc.), political or religious persecution, and other conditions that didn't allow them to grow, or even eke out a sustainable living. This familiar storyline will play out again as humanity gets busy beyond Earth.

Space science programs have a history of being highly productive. A very important part of their recipe for success is the scientific community's ability to stay focused on the big, important questions that challenge their various scientific disciplines. The space exploration and development community needs to devise its own approach to answering its big questions in a coordinated way. Before doing that, there needs to be agreement on what the big questions are. Perhaps they can be stated in this way:

- Can humans "live off the land" in space? If so, what integrated set of technical systems would be required to accomplish this? If not, how much can be achieved with robotic systems alone?

- Can expansive space operations consistently create value – scientific, economic, and societal – sufficient to justify the cost and risk? What are the priorities across the various investment options?

For human spaceflight projects, we have traditionally relied on destination-centric answers in response to questions of purpose. Using that approach, priorities are set and designs are locked in based on the narrow goal of getting from here to there and back. It is time to frame our future in a more sophisticated way which will develop cislunar space to bring benefits to Earth and prepare us to step outward to the rest of the solar system. This approach has been called Cislunar-Next. This alternative would de-emphasize sending humans to destinations beyond the Moon for now, and make in-space capabilities, infrastructure, and experience the top priorities. Some may perceive this as a "go slow" approach, but that would be a mischaracterization. In fact, if done properly, this would be the fast track to a purposeful, sustainable future in space. Significantly, it would be the best way to get the commercial sector on board as indispensable partners. Private investment interests will have skin in the game, and their partnership – perhaps even their leadership – will eventually propel the movement beyond cislunar space. If we rush to send humans to far-off destinations before achieving the industrialization of cislunar space, the commercial sector will participate primarily as government contractors, and will not be involved as a sustaining force.

Seeking A New Policy Recipe

Although current policy is increasingly capabilities-driven, it still clings to the traditional destination-driven approach. President Obama's National Space Policy of June 2010 directs NASA to aim for human visits to an asteroid by 2025 and to Mars orbit by the mid-2030s. As we have seen multiple times, a U.S. president has chosen to rely on the old destination-driven rationale to define the nation's human spaceflight goals. As in previous administrations, this approach was used because it is familiar, easy to understand, and seems to absolve the decision-makers of any further need to justify or expand on the vision. That includes explaining its underlying purpose and the benefits it brings to the nation and the world -- which are worth the cost and risk. Lacking such a well-formulated rationale, it would have been wise to leave out of the policy any mention of human missions beyond the Moon until their specific purpose and their place in a long-term strategy could become more firmly established.

The U.S. government's role in space development should be articulated in a long-range national policy with clearly defined approaches to managing the evolution of the civil space sector and facilitating the growth of the commercial space sector. For starters, we must recognize that space exploration and development will not evolve the way they did in the Cold War era. The U.S. government should not be expected – and in fact, is not able – to fund, build, and operate all the needed research projects, services, and infrastructure. The community of participants in research, operations, and funding needs to be enlarged. Many people are recognizing this, but it still needs to be better reflected in the principles and goals that are the essence of a good policy.

Space evolution in the 21st century will require a transition away from the way we've planned and prioritized our space activities in the past, so we need to start by articulating the principles that will guide actions under our modernized U.S. exploration and development policy:

- Space exploration and development shall be undertaken for the purposes of increasing scientific knowledge, improving stewardship of Earth, adding value to the national and global economy, enhancing international cooperation, and in general, extending human activity into the solar system for peaceful, beneficial purposes.

- Government-funded space infrastructure projects shall have applicability beyond a single mission or short-term series of missions.

- New operational capabilities and infrastructure created in U.S. government space development programs shall be designed for transfer, as early as possible, to operational entities in the U.S. government, private sector, or nonprofit sector.

- Operations beyond limited-duration science missions and engineering test projects shall not be assigned to NASA or other U.S. government research and development (R&D) organizations.

- U.S. government exploration and development missions will include humans when their presence is expected to yield cost-effective benefits or otherwise uniquely contribute to mission success and/or the national interest.

These principles establish a philosophy and work environment, which facilitates the concurrent evolution of exploration and development employing partnerships among government, nongovernment, and international players, each performing the roles most appropriate for them. The next step is to set ambitious but achievable goals, which progressively add space capabilities and contribute to global solutions.

The new goals must be more precise than in past policies, which typically have called for broad, ill-defined actions like advancement of U.S. interests and expansion of human activity into the solar system. But precise does not have to mean restrictive or lacking in ambition. In fact, the opposite is true: the goals must be both flexible and bold. They should be viewed in two different timeframes: short- to medium-term (2010s to 2030s) and long-term (2040s to the end of the century). Also, they should be recognized as goals that draw together the efforts of the whole nation, since this enterprise will demand more than can be achieved solely through government programs and investment.

Short-to medium-term goals should seek to develop enduring infrastructure, skill sets, and experience that will be essential for living, working, establishing communities, and creating value in the inner solar system. These goals are capabilities – not destinations – that will be essential for creating a spacefaring society that can expand its knowledge, economy, and sustainability. They include developing the technologies, processes, expertise, and infrastructure for:

- Utilizing the unique characteristics of space, such as microgravity, vacuum, high-intensity solar exposure, and isolation from Earth, to produce useful knowledge and products.

- Harvesting and processing of extraterrestrial materials and energy resources.

- Building progressively more sophisticated structures in Earth orbit and elsewhere in cislunar space.

- Building installations on the Moon, constructed to the greatest extent possible with local materials.

- Advancing space robotics to minimize the need for human presence in activities that are hazardous, remote, or are strong candidates for automation, and to provide direct assistance to humans where human involvement is required.

- The achievement of these goals should lead to the following long-term goals, starting around mid-century:

- Construction and operation of advanced structures that minimize their dependence on supply lines from Earth, designed for science, commerce, and other purposes.

- Aggregation of space structures into industrial parks at locations deemed valuable for their proximity to space resources, Lagrange points, or other attributes.

- Realization of significant contributions to the terrestrial economy through energy and manufactured products for use on Earth and in space.

None of these goals specify a planetary destination beyond cislunar space. Certainly, the Moon and near-Earth objects will be early destinations due to their close proximity and the broad range of contributions they can make to these goals. What comes next, and when it should come, should be driven by progress toward the goals, the rate of technological advance, the lessons of experience, and the availability of resources from all participants.

If the principles and goals suggested above require that ongoing operations be kept separate from research organizations, and that our top priorities are advanced knowledge and useful capabilities rather than specific destinations, we can begin to piece together a strategy which will put us on a path to our goals. At a minimum, such a strategy would require the following:

- Redirect NASA to focus more exclusively on R&D and shed its operational duties, other than limited-duration science and engineering mission operations that support its R&D programs.

- For operational and infrastructure components created by U.S. government space development programs, establish the identity and relationship of the intended system operator at the beginning of the program. The intended operator may be an operational entity of the U.S. government, the U.S. private sector, the U.S. nonprofit sector or an international consortium, but not an R&D organization.

- Set target dates for achieving capability milestones. Plans for reaching planetary destinations shall flow from the achievement of relevant capability milestones, not the other way around.

Principles and goals should be designed to endure, while the strategies and programs supporting them should be allowed to evolve. A document providing top-level policy guidance, such as a presidential directive or authorizing legislation should not get into specific programmatic decisions, which must be allowed sufficient flexibility.

The political and organizational difficulties of executing a significantly altered U.S. civil space policy should not be underestimated. Major components of NASA would be profoundly changed. The agency has become accustomed to designing and procuring its own space systems and operating its own launch facilities. This is a natural result of the fact that for much of the agency's history, there was no other entity willing and able to do all of this for it, especially in the case of human spaceflight. That is changing in the 21st century, and ultimately NASA must embrace that which it helped to create. If these new space services can do the job, this will mark a new level of spacefaring maturity spawned by NASA's stimulation of technologies and markets. This is one of the big payoffs we've been waiting for, so NASA should feel proud rather than threatened.

This policy proposal may seem to be quite a stretch, given the short planning horizon that's typical in our political system. But there is precedent for taking a longer view. For example, when the U.S. Navy plans for the future of its fleet, especially complex big-ticket requirements like an aircraft carrier group, it must think through the entire life cycle – at least 30 to 40 years – including development, deployment, decades of operation, and retirement. That means today's Navy planners are working in a timeframe which stretches through mid-century. Should we expect any less from the planners of our space infrastructure?

The Three Stages of Space Exploration And Development

The spacefaring goals and strategies outlined above are designed to support three stages of space development. Stage One, where we have been for the past half century, uses space as a training ground for technical systems and operational experience, yielding useful applications that employ the vantage point of space. In Stage Two, we will turn cislunar space into an industrial park, going beyond remote imaging and signal relays to begin generating value from extraterrestrial resources. Stage Three begins with the sustainable, permanent settlement of cislunar space and the initial expansion of human activity out to the rest of the solar system. These stages cannot be reshuffled or skipped over if we want to create an enterprise with lasting value.

We seem to be a long way from reaching Stage Two. How long it takes us to get there is dependent on good choices, sustained commitment, and a reasonable amount of good luck. At this point in our development, it is critical that we make the right strategic choices as we decide what will be the "Next Great Thing" in space.

Though less dramatic than headline-grabbing milestones like "the first humans on Planet X," creating Stage Two infrastructure and industries will challenge us to build new and better space capabilities, expand the frontiers of science, bring direct benefits to Earth, and eventually enable us to achieve Stage Three. The secondary benefits, including technology spinoffs and inspiration of our youth, are likely to be realized in great abundance, beyond anything we achieved in the aftermath of Apollo. But the enterprise must be justified by its primary benefits.

Initial Steps to Stage Two

We already have an assortment of Earth-to-orbit launchers and satellite fleets that perform a variety of valuable functions for our security, our economy, and our scientific pursuits. All of this has made an immense contribution to improving the human condition, but it's just the beginning. The next step will require more than evolutionary improvement of the things we are already doing, which mostly consist of transmitting electromagnetic signals back and forth. We will need to be able to harvest and process raw materials and energy in space. We will need to build things, large and small, in orbital space and on the Moon. We will need laboratories, manufacturing facilities, and habitats. We will need the means to efficiently transport people, cargo, and automated systems throughout cislunar space. Vital to all of this is proximity operations, the ability to rendezvous, dock, and otherwise conduct activities involving separate spacecraft interacting in close proximity or direct physical contact.

Identifying proximity "ops" as the key to our future in space may seem odd since we've been doing rendezvous and docking of spacecraft since the 1960s, and it's a routine part of space station operations today. However, the bulk of this activity has been associated with high-profile human missions in low Earth orbit (LEO), and in lunar orbit during the Apollo era. Today, there are no operational manned or robotic systems that can perform rendezvous, capture, repair, refueling, reboost, and retrieval of orbiting payloads throughout cislunar space. Since the retirement of the space shuttle, there is not even an operational system that can do these things in LEO. Such systems are needed if space development is to advance beyond its current stage. This will be true across the civil, commercial, and national security space sectors.

Technical and funding challenges have limited our progress in this area, but they aren't the only concerns. Proximity ops are a touchy subject in many parts of the space community. Despite their potential benefits, commercial operators and insurers worry about the risk of collision or other "oops" moments that could damage a spacecraft or send it spinning. This could permanently end a lucrative revenue stream for a commercial satellite and cause a debris hazard that the operator would be responsible for.

National security operators share the same concerns about the risk of service interruption or permanent damage. In addition, they would rather not see a proliferation of capabilities for getting up close and personal with their satellites. In fact, they would prefer that no one else knew exactly where those satellites were, or even the fact that they existed.

Anti-satellite weapons (ASATs), which can be seen as a specialized application of proximity ops, have prompted concerns since the beginning of the space age. The more we improve guidance, navigation, and control systems – and the more those systems proliferate – the greater the concern becomes. This already was evident in the early post-Apollo years, when the Soviet Union was testing a co-orbital explosive ASAT.

It seems abundantly clear that if we have any plans of advancing the development of space, we will need to overcome our fear of proximity ops. To prohibit or excessively restrict these capabilities just because they pose a risk of accidents or could function as weapons would mean halting further space development. The global community must accept proximity ops and establish behavioral norms that dispel fears and tensions. This is how we deal with other technologies such as aircraft, ships, ground vehicles, and cell phones. All of these have been used in crime, terrorism, and war, but it would be absurdly counterproductive to ban them.

With all of this in mind, consider next what the work program would include if we decided to pursue Cislunar-Next. What capabilities do we need to demonstrate in the Earth-Moon system that would be valuable elements of a space infrastructure? What are the high-priority proof-of-concept projects?

Setting Priorities

Some research and development areas will be essential regardless of the approach chosen for human spaceflight. For example, human sustainment systems will require continuous improvement in the knowledge and techniques for dealing with the physiological and psychological stresses of long-duration missions. Life support systems need to become more reliable, lower maintenance, and less dependent on frequent resupply. This is an area that is already getting a good deal of attention aboard the International Space Station (ISS) and in some Earth-bound studies. But we lack the facilities to do all that is needed.

The crew conditions being studied on ISS are confined to six-month stays (and one full-year mission planned for 2015) in zero-gravity in low Earth orbit, giving us a very limited sample of what we'll encounter as we move outward. This is far short of the time needed for any interplanetary journeys or extended cislunar missions, and provides minimal ability to prepare us for the radiation levels that confront us when we go beyond the relatively benign environment of low Earth orbit. Radiation exposure may be the greatest potential showstopper to long-duration spaceflight and habitation unless adequate mitigation measures can be developed.

Another ISS limitation is that it tells us nothing about how to function on a planetary surface with gravity that is a fraction of Earth's. The weightless environment of the ISS needs to be supplemented by a variable gravity facility that can simulate planetary gravity environments that we expect to encounter, such as one-sixth-g on the Moon or one-third-g on Mars. Also, the facility can help determine if spinning spacecraft make sense for long flights, and if so, at what gravity level. A one-g environment may not be necessary to maintain good health and full functionality, and there are technical advantages to designing a spacecraft for lower spin rates.

Some of the big questions that desperately need answers may receive cursory attention if we choose a spaceflight approach that only looks beyond the Moon and is overly eager to put human footprints on alien worlds. Under Cislunar-Next, developing the technologies and experience that enable efficient, sustainable, expandable space operations would be top priority. Let's consider some of the essentials for Cislunar-Next that might receive inadequate attention under other approaches that simply aim at stepping-stones to Mars and other solar system destinations.

On-orbit servicing

If we are serious about living and working in space for the long haul, we are not going to discard our hardware every time it breaks down or runs out of juice. We are going to learn how to refill its tank, replace its gaskets, give it a tune-up, extend its life, and upgrade its capabilities. This has to become routine, unlike the elaborate and expensive Hubble Space Telescope repair missions. As much as possible, we will do the job with automated or tele-operated robots. Tele-operation is facilitated by the fact that everything in cislunar space is less than a light-second away.

Standardization

If retrieval, repair, and refueling of space hardware is to occur, it will be enabled and assisted by standardization. Manufacturers will need to redesign their space hardware to be serviced by robots using common interfaces. This should not be especially difficult, since only a small number of organizations worldwide routinely produce large satellites, and they are already doing standardization to integrate their commercial payloads with multiple launch vehicles from several countries. By comparison, it would seem a simple matter to settle on standard grappling fixtures so that service vehicles can capture satellites safely and efficiently. Also needed are standard ports for fuel and other fluids, electric power, and data transfer. Replacement of old or malfunctioning parts could be done with modular components. Once these standards are in place, they can be carried over to modular assembly of large platforms.

Inter-orbital transportation

When NASA first envisioned the Space Transportation System in the 1970s, it was going to be more than just the shuttle. Another element was called the Orbital Transfer Vehicle; a reusable upper stage (or "space tug" as it was also called) that would have taken payloads from the shuttle's orbit to higher orbits. In the coming decades, in-space transportation needs to have a renaissance comparable to the experience of automobiles, ships, and aircraft in the 20th century. This will produce a wide variety of craft that are sized and specialized for particular tasks. Just as cars, trucks, ships, and aircraft come in an assortment of shapes and sizes, so will future space vehicles that travel between LEO, GEO, lunar orbit, and Lagrange points. They will drop off and retrieve many kinds of payloads, and will carry robots and humans to locations where they're needed.

Fuel storage

NASA has considered the possibility of using an on-orbit fuel depot in low Earth orbit that would fill up the final rocket stages of missions bound for the Moon or points beyond. It is difficult to see how NASA could justify such a facility if it served only the agency's own deep space missions flying an average of less than once per year (as NASA's own study assumed). Alternatively, orbiting fuel depots developed for Cislunar-Next would not necessarily be in LEO, nor would they always get their fuel supplies from Earth. They would be located where the action is, be it in LEO, GEO, lunar orbit, or Lagrange points. Some of their supplies could come from the hydrogen and oxygen extracted from lunar or asteroidal ice. Their best customers would come from the inter-orbital traffic throughout cislunar space (for example, satellite servicing bots and reusable orbital transfer stages) with less-frequent visits from deep space missions needing a fill-up on their way out. The customer list would encompass the world's spacefarers, public and private – not just a single agency from a single country. This is another place where standardization would be essential. Just like terrestrial gas stations, all manner of space motorists should be able to pull up to the pump, swipe their credit card, and neatly fit the dispenser nozzle into their tank.

Materials processing in space (MPS)

An important component of the space economy will be microgravity materials processing. The reaction of some people with long memories may be, "Been there, done that, didn't work." But that quick dismissal would be a mistake. Since the 1980s, there has been plenty of attention devoted to MPS in attempts to discover unique properties and take advantage of processes and conditions not available on Earth. However, there has not been nearly enough on-orbit research. Access to lab space on orbit has been extremely hard to obtain, and very expensive to use, so the basic research phase of this activity has stretched out with no apparent end in sight. We still have not answered questions about which processes result in useful products, how those processes might be scaled up to industrial production levels, and whether any of this can be turned into a business plan with a happy ending. It is vital that we answer these questions. There will not be much of an economic future in space if all materials processing and manufacturing has to be done back on Earth.

Extraterrestrial resources

Science fiction writers and real-world space planners have been talking for decades about in-situ resource extraction and utilization, but we still do not have definitive answers to first-order questions. Can we mine the Moon and asteroids for minerals? How would terrestrial mining methods need to be modified for the task? Should materials be refined on site, or in a separate facility? What kinds of final products will use these materials? Will the products only be used in space, or will they be marketable on Earth?

Much has been made of the strong evidence that large deposits of water ice exist in permanently shadowed craters near the poles of the Moon. That is great news, but those deposits still need to be located precisely and their extent has to be estimated more accurately. Then we need to figure out how to "mine" the ice, which will require specially designed systems that can function in extremely low temperatures. Once extracted, the ice must be transported to a facility for processing, to turn it into potable water or to separate the hydrogen and oxygen components to supply fuel, oxidizer, or breathable oxygen. All of this must be demonstrated before we can count on lunar ice as a critical element in the cislunar infrastructure.

Energy collection and distribution

Adequate power needs to be available at widely dispersed locations. What should be the balance between solar, nuclear, and fuel cell power sources? What particular system design is best in each of these categories? We will not know the answers to these questions until fully engaged in Cislunar-Next.

Long before we feel the need to import minerals to Earth from space, we may find ourselves looking skyward for a source of abundant clean energy. Solar power platforms can collect the Sun's energy and beam it to where it is needed, whether that's other orbiting platforms, lunar outposts, or the surface of the Earth. But we have yet to conduct a pilot project to demonstrate this in space, which must be followed up by efforts to scale it up to industrial size. If we are developing techniques for lunar materials extraction and processing at the same time, the solar platforms could be the driver for exploring how much we can build in space with materials that do not have to be lifted up from Earth.

Other in-space utilities

If operations throughout cislunar space become routine, there will be a need for dedicated communications and navigation services similar to the ones we use on Earth and in LEO. Existing services are all aimed at serving Earth, so additional systems are needed to serve other parts of cislunar space. That does not mean the Moon needs a 30-satellite constellation for navigation, or that lunar orbit should be filled with communications satellites comparable to those in GEO. But, operations on the scale being discussed here can no longer depend on research facilities like NASA's Deep Space Network to provide all that is needed. Whatever architecture is chosen to provide "comm and nav" to cislunar space, it should be designed to operate as transparently as possible to its users without routing everything through Earth and a room full of mission controllers.

Another essential utility is space weather monitoring. Human crews living and working in high orbits or on the Moon need timely warnings and analyses of solar activities that could have dire effects on their health and technical systems. Ideally, they should have real-time links to the warning systems to avoid any delays in alerts from Earth-bound observers. Future human activities spread across cislunar space may not have the luxury of around-the-clock monitoring by teams of technicians, as the ISS does today.

This discussion of priorities is not intended to be an exhaustive list of the things we can or should do in cislunar space, but it does show that even after a half-century of spaceflight, there are plenty of things we do not know and cannot do in our own celestial backyard. The point is not to pave the way for countless thousands of people to work in space as quickly as possible; in fact, the more we can do with automated systems, the better. The point is to identify the hurdles that need to be overcome to vastly increase humanity's resources, capabilities, and knowledge.

No One Said It Would Be Easy

The challenge for the space community is to progress from concepts to operational systems – in fact, to an infrastructure of integrated systems – in an environment that is resource-constrained and faces the countervailing forces of short-term thinking, partisanship, and parochialism. These forces have been around throughout the space age and have always been influential. However, in recent decades, they have increased the danger that the vital enterprise of space exploration and development will become little more than a hobby shop for builders of rockets and spacecraft and a jobs program for a handful of congressional districts. If that is the case, then the United States has lost something of great value: one of its most potent tools for building a better future.

The U.S. will probably always retain the ability to build space hardware, but that's not the same as building a future. We can do a better job of giving a purpose (or better yet, multiple purposes) to the nation's evolving space systems. We can carry out missions in support of Cislunar-Next. We can build things, like microgravity laboratories, manufacturing facilities, lunar outposts, and solar power satellites. We can create or enhance capabilities, like satellite servicing and the harvesting of extraterrestrial energy and material resources. In other words, we can do things that generate scientific, economic, and societal value, bringing quantitative and qualitative benefits to Earth and justifying the continued exploration and development of space. The cost of *not* doing these things may be unaffordable.

In order to move to the next level of maturity in our quest to become Stage Two spacefarers in the coming decades, we need to embrace what's been created so far in the space age – much of it through NASA's influence – by taking advantage of growing cross-sector abilities. Offload operational and routine functions to the private sector and other entities that are increasingly equipped to handle them, and let NASA focus on the scientific research and technology development activities at which it excels, and which otherwise would receive inadequate attention. Our metrics for success should not be based on how quickly we get to Mars or how many people we have living in space; rather, we should be measuring how much we're gaining in capabilities and knowledge, leading to increased prosperity, global solutions, and discovery.

Advancing U.S. Geopolitical and International Interests in Space

Scott D. Pace

Whether it was demonstrating America's peaceful intention for its space program, beating the Soviets to the Moon, beginning Space Station Freedom to better cement allied interests in space, welcoming Russia into the community of liberal democracies after it threw off the yoke of communism by merging the U.S. and Russian human spaceflight programs in the International Space Station (ISS), or working with other countries to increase mankind's knowledge of earth, space, planetary, life, and microgravity sciences for mutual benefit, the civil space program has long contributed to America's geopolitical interests (and, for that matter, the interests of its partners). By offering pragmatic and attractive opportunities for space cooperation to other countries, the United States encourages them to draw closer to it, its space and international agenda, its interests, and its goals. Thus, the space program has geopolitical value: it can contribute to the pursuit of American interests.

Space activities play critical roles in U.S. national security, economic growth, and scientific achievements. For example, the Global Positioning System (GPS) is an integral part of several critical infrastructures and enables functions ranging from survey and construction, to farming, finance, and air traffic management – not to mention supporting U.S. military forces worldwide. The International Space Station represents a unique, collaborative partnership between the United States, Europe, Canada, Japan, and Russia.

At the same time, new threats to U.S. space activities have emerged, threats that are different from those of the Cold War and which have their own distinct dynamics. Some threats are intentional and others are not. In some cases, threats come from a known nation state while, in others, it is impossible to identify the source of a threat due to a lack of full "space situational awareness," that is, knowledge about what is happening in space, not to mention who is doing what. For example, China tested a high altitude anti-satellite weapon (ASAT) against one of its old weather satellites in 2007 without first warning the international community. This test created tens of thousands of pieces of orbital debris and increased the risk of collision and damage to many satellites operating in low Earth orbit, including the International Space Station, for many years. In 2009, there was an accidental collision over the Arctic between a defunct Russian Kosmos communications satellite and an active Iridium communications satellite that added even more orbital debris to low Earth orbit. North Korea has continued developing ballistic missile capabilities under the guise of peaceful space launches. Iran continues to jam commercial satellite broadcasts in order to prevent foreign reports of domestic unrest from reaching its population. There have been reports of attempts at unauthorized communications access to U.S. civil scientific satellites, e.g., Terra and Landsat in 2007 and 2008, but the source of these attempts has not been confirmed. Threats to the sustainability of space activities today come not from a single superpower, but from a much more diverse group of actors whose motivations can range from deliberate to ambiguous and even accidental.

The global space community is a dynamic one with new capabilities and new entrants. Europe is building its own version of GPS, titled

"Galileo," and has long been a leading supplier of international commercial launch services with its Ariane family of launch vehicles. China has flown several astronauts, becoming only the third country with independent human access to space. Moreover, it is constructing a space laboratory and has demonstrated unmanned rendezvous and docking operations in preparation for a fully manned space station in 2020 – about the time the International Space Station may end its operations. Japan has announced plans to sell radar satellites to Vietnam while South Korea seeks to sell an optical imaging satellite to the United Arab Emirates. Brazil and China are continuing many years of space cooperation in remote sensing while India and South Africa are close to concluding their own space cooperation agreement. All of these countries recognize that space capabilities are important for both practical and symbolic reasons and that these capabilities are intrinsically "dual-use" in that civil, security, and commercial applications are based on similar skills and technologies.

At home, the U.S. civil space program has dealt with multiple changes in policies, program, and budgets in recent years. These developments are the result of changes in Administration policy, increasingly constrained and volatile budgets, the completion of International Space Station construction, and the subsequent end of the Shuttle program. The nation's national security space programs are also under great stress due to technical shortfalls, schedule delays and cost overruns. As with civil space, fixing these programs will be increasingly difficult in the increasingly constrained budget environment.

The technical and budgetary risks facing civil and national security space programs are merely the more visible symptoms of deeper policy and management disconnects between the White House and Congress. These disconnects affect U.S. national security and foreign policy interests as well as scientific and economic objectives and reflect a lack of coherence in the oversight and execution of U.S. space policy. Disconnects and disarray are not inevitable and can be resolved through the adoption of what can be termed a more "geopolitical" approach to creating a stable and sustainable foundation that can advance U.S. national interests and the interests of our friends and allies. International space cooperation, space commerce, and international space security discussions could be used to reinforce each other in ways that would improve the sustainability and security of all U.S. space activities as well as contribute to U.S. geopolitical goals beyond space.

Flaws in U.S. National Space Policy and the National Security Space Strategy

The current U.S. National Space Policy, a comprehensive document that addresses the full range of U.S. interests in space, was released in June 2010. The policy continues many long-standing principles, such as the right of all nations to engage in the peaceful uses of outer space, recognition of the inherent right of self-defense, and that purposeful interference with space systems is an infringement of a nation's rights. The policy states that the United States "recognizes the need for stability in the space environment" and that it will pursue "bilateral and multilateral transparency and confidence building measures to encourage responsible actions in space."

The policy made some important changes compared to the 2006 National Space Policy, notably with respect to arms control. The 2010 policy does not categorically reject space-related arms control that would constrain U.S. space activities, but states that any such agreements would have to be "equitable, effectively verifiable, and enhance the national security of the United States and its allies." This is a traditional policy formulation that was also used during the Reagan Administration. There is nothing in either the 2006 or 2010 policies, however, about actively pursuing new international treaties, creating legal norms, or characterizing space as a "global commons" or being part of the "common heritage of all mankind" – legally ambiguous terms that may create unforeseen constraints on U.S. freedom of action in space.

The general coherence on the national security and foreign policy side is not matched in the section of the 2010 National Space Policy dealing with civil space exploration. The policy says that the NASA Administrator shall "set far-reaching exploration milestones. By 2025, begin crewed missions beyond the moon, including sending humans to an asteroid." Unlike the carefully crafted text elsewhere in the policy, this section appears to have been taken directly from an April 15, 2010 speech by President Obama at the Kennedy Space Center in Florida. Subsequent technical work has shown that there are few scientifically attractive, technically feasible asteroids that can be reached on this schedule. Even worse, the international space community, which had been shifting attention to the Moon in anticipation of that being the next U.S. focus of exploration beyond low Earth orbit, felt blindsided. Countries in Asia, such as Japan, India, China and South Korea, saw the Moon as a challenging but feasible destination for robotic exploration and a practical focus for human space exploration. The asteroid mission was, perhaps unintentionally, taken as a sign that the United States was not interested in broad international cooperation but would focus on partnerships with the most capable countries, such as Russia and perhaps Europe. As a result, spacefaring countries are increasingly making their own space exploration plans separate from the United States.

The perception that the next steps in human space exploration would be too difficult to allow meaningful participation by most spacefaring countries undercut international support for human space exploration more generally. The lack of U.S. support for a program to return to the Moon made it difficult for advocates of human space exploration in Europe, Japan, India, and elsewhere to gain funding for any efforts beyond the International Space Station. The ISS is itself under budget pressure to justify its construction and on-going operations costs, a task made more difficult by the lack of a clear direction for human space exploration beyond low Earth orbit. The lack of international leadership by the United States provides an opportunity for rising spacefaring countries such as China to play a greater role in the future. If China is able to offer pragmatic opportunities for space cooperation on its own space station or as part of efforts to send humans to the Moon, other countries will likely find it attractive to forge closer relationships with China. A shift in international space influence away from the United States and toward China would have the potential to impact a wide range of U.S. national security and foreign policy interests in space.

While the U.S. National Space Policy is supposed to be authoritative guidance, a subsequent document on national security space activities deviated from it in some key respects. The National Security Space Strategy was released in January 2011 as a report to Congress and is intended to provide direction to the national security space community in planning, programming, acquisition, operations, and analyses. Despite being signed by the Secretary of Defense and the Director of National Intelligence, it places a major emphasis on diplomatic activities – a responsibility of the Department of State – as well as dual-use capabilities that are promoted and regulated by the Departments of Commerce, Transportation, State, and the Federal Communications Commission (FCC). The implementation of the strategy says, in part, that (emphasis added):

"We seek to address congestion by *establishing norms*, enhancing space situational awareness, and fostering greater transparency and information sharing."

"We seek to address the contested environment with a *multilayered deterrence* approach. We will support establishing international norms and transparency and confidence-building measures in space, primarily to promote spaceflight safety but also to dissuade and impose international costs on aggressive behavior."

The problem with the phrase about establishing norms is that it goes beyond the terms of the National Space Policy. Furthermore, it presupposes that there will be some authority by which the norms are established and that the United States will be bound along with other nations. Many harmful activities such as the intentional creation of long-lived orbital debris and intentional satellite jamming are already contrary to international law, notably the Outer Space Treaty and the Constitution of the International Telecommunication Union. Yet there has been little in the way of sanctions save for international complaints. A more useful statement might have been one about promoting compliance with existing international laws and agreements.

Unfortunately, the Defense Department also uses the legally problematic term "global commons" with respect to space in the most recent Quadrennial Defense Review (February 2010). This term applies to the high seas and the air above them, but is not yet accepted internationally or even officially by the United States. Whether intentional or not, use of the terms "norms" and "global commons" sends mixed messages to international audiences about the U.S. view of space, despite stated Administration desires to reduce miscommunication.

Subsequent statements by Defense Department officials expanded on this new concept of "multilayer" deterrence with respect to U.S. space systems. In an April 13, 2011 speech to the 27th National Space Symposium, the Deputy Assistant Secretary of Defense, Greg Schulte, outlined four layers of deterrence with the first being the "establishment of norms of responsible behavior," and the second being the establishment of "international coalitions" such that an attack on the space capabilities of one country would inevitably be an attack on the capabilities of other countries. The third layer of deterrence would be to "increase the resilience and capacity" of space systems to operate in a "degraded environment," and the fourth layer would be the "readiness and capability to respond in self-defense, and not necessarily in space," as a means of complicating any adversary's plans to attack space assets. The latter two layers are arguably part of traditional deterrence theory in which an opponent is deterred through fear of retaliation or denial of attack objections. The first two layers may seem plausible, but in practice they represent a gross oversimplification of possible foreign reactions, including those of our allies. Norms of responsible behavior are unlikely to be a serious consideration to an adversary contemplating an attack on U.S. space systems. Under threat of attack or denial of crucial space systems, other countries may move toward neutrality in a crisis. In the longer term, they may accelerate the acquisition of independent capabilities rather than be unwillingly "entangled" with the United States.

China and Russia have for many years advocated an international treaty barring space weapons as well as the use or the threat of the use of force against space objects. They have introduced a draft treaty at the UN Conference on Disarmament as part of deliberations on the "prevention of an arms race in outer space." The United States has consistently opposed such a treaty as unnecessary, unverifiable, and not in the interests of the

United States and its allies. A major flaw in the draft treaty is the difficulty in defining just what a space weapon is; even if defined, the Chinese-Russian text leaves out ground-based systems such as interceptors and lasers. Consideration of a verifiable and more meaningful agreement, based on behavior, to ban the intentional creation of long-lived orbital debris has not gained much traction due to the impasse over the Chinese-Russian proposal.

How the United States can move beyond the current stalemate in discussions of international space security and reclaim a stronger leadership position among other spacefaring countries will be addressed later. First, we need to address the role of human space exploration, which once had a clear strategic rationale in the race to the Moon, but which has been the subject of prolonged debate ever since.

Finding a Strategic Approach to Human Space Exploration[8]

Unmanned space activities, whether for science, commerce, or national security, face severe cost, schedule, and performance pressures. The problems faced by unmanned programs are qualitatively different from those faced by human space exploration. Scientific, commercial, and national security space programs have relatively clear rationales and processes for deciding among competing priorities. There is no debate in the United States about whether, for example, to have a planetary science program, but rather what level of effort is affordable and executable. The debates over unmanned space activities represent programmatic, not existential, questions.

In contrast, there is an on-going debate over whether and what kind of human space exploration effort the United States should have. While

---

[8] Portions of this chapter are derived from my testimony on "NASA's Strategic Direction" before the House Committee on Science, Space, and Technology, December 12, 2012.

many supporters of human space flight see such efforts as "inevitable" or "part of our destiny," those views are not widely enough held to ensure stable political support. At the same time, there is a level of support for the symbolism of human space flight and a sense that it may have longer-term practical value that make U.S. political leaders reluctant to cancel such efforts or to be seen as supporting such an action. Human spaceflight (if not pure exploration) may one day become a self-sustaining commercial activity but that day has not yet come. As argued below, the most immediate benefit of human space flight today, aside from technical and scientific achievements, lies in its contribution to U.S. prestige and geopolitical leadership.

There are many diverse reasons individuals may have for supporting human space flight along with many different activities that could constitute an on-going human space flight effort, e.g., space tourism, landing on Mars, exploiting space resources, etc. Aside from an Apollo-like political crisis, which seems unlikely to reoccur, there are three major alternative strategic approaches the United States might take toward human space exploration: *Capability-driven, Question-driven,* and *Geopolitically-driven*.

Capability-driven

The current U.S. approach to human space exploration is officially described as "capability driven":

> NASA's human space exploration strategy focuses on capabilities that enable exploration of multiple destinations. This capability-driven approach is based on a set of core evolving capabilities that can be leveraged or reused, instead of specialized, destination-specific hardware. This approach is designed to be robust, affordable, sustainable, and flexible, preparing NASA to explore

a range of destinations and enabling increasingly complex missions.[9]

This approach does not focus on a specific destination, question, or purpose for human space flight, but rather seeks to keep a range of options open while deferring decisions on specific architectures and rationales. In a budget constrained environment without any specific political or economic rationale, such an approach avoids both the need to make a decision to cancel human space flight, or, if it is not to be cancelled, the need to specify what it is that human space flight should accomplish.

This is not the first time the United States has taken this approach. In the aftermath of the Apollo program, the Nixon Administration did not want to cancel human space flight but neither did it want to continue the costs and risk of human missions to the Moon and eventually Mars. In 1970, while the lunar landings were still underway, President Nixon said:

> We must realize that space activities will be a part of our lives for the rest of time. We must think of them as part of a continuing process-one which will go on day in and day out, year in and year out -- and not as a series of separate leaps, each requiring a massive concentration of energy and will and accomplished on a crash timetable.... We must also realize that space expenditures must take their proper place within a rigorous system of national priorities.[10]

NASA Administrator James Fletcher explained the 1972 decision to build the Space Shuttle in a similar, low-key fashion:

There are four main reasons why the Space Shuttle is important and is the right step in manned space flight and the U.S. space program.

---

9 NASA, "Voyages: Charting the Course for Sustainable Human Space Exploration," Washington, D.C., June 7, 2012, http://www.nasa.gov/exploration/whyweexplore/voyages-report.html.
10 T.A. Heppenheimer, *The Space Shuttle Decision* (Washington, D.C.: NASA, 1999), accessed at http://history.nasa.gov/SP-4221/ch9.htm. See Chapter 9, "Nixon's Decision."

1. The Shuttle is the only meaningful new manned space program which can be accomplished on a modest budget;

2. It is needed to make space operations less complex and less costly;

3. It is needed to do useful things, and

4. It will encourage greater international participation in space flight.[11]

In essence, NASA would develop a human space flight capability that would continue to enable the United States to send humans into space, be more affordable, and hopefully accomplish useful tasks still to be determined. The Obama Administration's current approach is arguably similar to that taken by the Nixon Administration in the early 1970s.

Question-driven

An alternative strategic approach is to take an intentionally question-driven approach and pose questions or grand challenges to be addressed by human space exploration efforts – or at least those efforts that rely on public resources. In this approach, a program of human space exploration is more than a series of spectacular engineering demonstrations – as in the case of Apollo – but a means of answering questions important to society.

After gaining foundational capabilities like space transportation, communications, navigation, and power, an exploration program could look to ways to use in-situ resources, create new resupply methods, and commercial partnerships. This could help move debates beyond "robots versus humans" or "Moon versus Mars" or "Science versus Exploration" to a more question-driven, mission-focused series of decisions.

---

11 NASA, Statement by Dr. James C. Fletcher, NASA Administrator, January 5, 1972, accessed at http://history.nasa.gov/stsnixon.htm

Just as the *Challenger* accident led to questioning whether human life should be placed at risk in launching satellites that could be carried by an unmanned rocket, so the *Columbia* accident led to asking for what purposes, if any, was risking human life worthwhile. The Columbia Accident Investigation Board (CAIB) concluded that the nation should continue a program of human space flight, eventually moving beyond Earth orbit. Although not stated explicitly, the implication was that if the nation were to continue to place human life at risk, staying in low Earth orbit was an insufficient goal to justify such risks.

For those who believe that human expansion into the solar system should be an important part of what the United States does as a nation, abandoning human space flight completely or even staying in low Earth orbit would be unacceptable. However, there are many who do not share the same feeling about the priority of human space flight to the nation, and it would be realistic to squarely acknowledge that uncertainty. The original decision to go to the Moon was an answer to President Kennedy's question on whether the United States had a chance of surpassing the Soviet Union in any area of space achievement. The change in payload policy after *Challenger* was an answer to the question of whether it was justifiable to risk humans for satellite deployments. After *Columbia*, the CAIB recommendation to eventually go beyond low Earth orbit was an answer to the question of whether humans should be in space at all.

Today, what is the question for which the human exploration of space is the answer? Such a question could be, "Does humanity have a future beyond the Earth?" Either a yes or a no answer would have profound implications. Addressing this question quickly leads to two sub-questions: can humans "live off the land" away from Earth, and is there any economic justification for human activities off the Earth?[12] If the answer to

---

[12] Harry L. Shipman, *Humans in Space: 21st Century Frontiers* (New York: Plenum Press, 1989).

both questions is yes, then there will be space settlements. If the answer to both questions is no, then space is akin to Mount Everest – a place where explorers and tourists might visit but of no greater significance. If humans can live off-planet, but there is nothing economically useful to do, then lunar and Martian outposts will, at best, be similar to those found in Antarctica. If humans cannot live off-planet, but there is some useful economic activity to perform, then those outposts become like remote oil platforms. Each of these scenarios represents a radically different human future in space. While individuals might have beliefs or hopes for one of them, it is unknown which answer will turn out to be true. The answer can only be found by actual experience and new information.

The science community has used the productive practice of posing simple but profound questions to shape and guide the implementation of research strategies. To ask "is there life elsewhere in the universe?" leads to questions of whether there is life elsewhere in the solar system, the search for water on Mars, and missions exploring for water and signs of life in particular locations. These questions shape the design and execution of space missions. The human space flight community could benefit from adopting similar practices to design and prioritize its missions. In this vein, consideration should be given to a routine survey that sets metrics and assesses progress in (or lack of) human spaceflight and reviews priorities on a ten year time scale as done for scientific fields. For example, priority could be given to answering such questions as:

- Can humans operate effectively away from Earth for long periods of time?
- Can we utilize local resources to lower reliance on materials from Earth?

- Are self-sustaining commercial activities (requiring direct or close human involvement) in space possible?

Such routine reviews could also improve the stability of human spaceflight efforts across Administration transitions. If the United States could shift away from existential debates on whether or not to have a human space exploration effort, it could use open, enduring questions to guide programmatic decisions for an affordable and effective human spaceflight effort.

Geopolitically-driven

The third strategic approach is the most historically common for the United States, a human space exploration effort driven by geopolitical interests and objectives. The United States undertook the Apollo program in the 1960s to beat the Soviet Union to the Moon as part of a global competition for Cold War prestige. The Apollo- Soyuz program symbolized a brief period of détente in the 1970s. The Space Station program was established in the 1980s, in part, to bring the developing space capabilities of Europe and Japan closer to the United States and to strengthen anti-Soviet alliances. Russia was invited to join a restructured International Space Station in the 1990s to symbolize a new post-Cold War, post-Soviet relationship with Russia. What might be the geopolitical rationale for the next steps in human space exploration?

It is well recognized that many of today's most important geopolitical challenges and opportunities lie in Asia. States under UN sanction, (Iran and North Korea), are seeking to develop ICBM capabilities under the guise of space launch programs. China, India, and South Korea are demonstrating increasingly sophisticated space capabilities that serve both civil and military purposes. Examples of these capabilities include satellite communications, environmental monitoring, space-based navigation, and scientific research. Unlike Europe, there are no established frameworks for peaceful space cooperation across Asia. In fact, the region can be characterized as containing several "hostile dyads" such as India-China, North Korea-South Korea, and China and its neighbors around the South China Sea.[13] The United States has better relations with almost all of these countries than many of them have with each other.

Asian space agencies have shown a common interest in lunar missions as the logical next step beyond low Earth orbit. Such missions are seen as ambitious, but achievable and thus more practical than missions to Mars and more distant locations. They offer an opportunity for emerging and established spacefaring countries to advance their capabilities without taking on the political risks of a competitive race with each other. A multinational program to explore the Moon, as a first step, would be a symbolic and practical means of creating a broader international framework for space cooperation. At the same time, the geopolitical benefits of improving intra-Asian relations and U.S. engagement could support more ambitious space exploration efforts than science alone might justify.

Integrating National Interests in Space

From the beginning of the Space Age, space activities have been "tools" of both hard and soft power for participating nations. Hard power is represented by alliances, military capabilities, and economic strength that can compel and pay others to do what we desire. Cultural, diplomatic, and institutional forces are aspects of soft power by which we are able to persuade others to do what we desire. In seeking to advance international space security interests, the soft-power influence brought about by leadership in civil and commercial space activities must be considered. Countries lacking a stake in a stable, peaceful space environment

---

13 James Clay Moltz, *Asia's Space Race* (New York: Columbia University Press, 2011).

are unlikely to be supportive of U.S. and allied space security concerns. It is not that those countries will oppose U.S. security goals, but that they will not see the relevance to their own needs and interests. As an example, international interest in mitigating orbital debris has grown as more countries have realized the threat such debris can pose to space systems they rely on and to their citizens working in space.

Potential international partners have been confused by a lack of clear U.S. space goals and priorities. Looking beyond the International Space Station, they have not seen opportunities for engagement other than in individual scientific collaborations. As one European space agency head put it, "there is lots of cooperation with Europeans, just not with Europe."[14] The International Space Station is the only example of strategic, as opposed to opportunistic, cooperation with Europe at present. It should go without saying that the United States should be in the position of advocating and leading new strategic initiatives, rather than merely responding to those of others.

Now that construction of the International Space Station has been completed, the priority of all the partners is rightly on utilization. If the current international partners do not see the ISS as a success, it is difficult to imagine international support for new human space exploration efforts. However, since major space projects take so long to implement, it is appropriate to work now on what should come after the Station – even if the Station's end date is not certain. It is generally assumed that individual nations will not undertake human space exploration beyond Earth orbit. So, it makes sense to ask potential international partners what they are capable of and interested in doing. In this regard, human missions to asteroids or Mars are beyond the practical capabilities of almost all potential partners but can still serve as long-term goals.

---

14 Personal communication.

If there is to be a serious effort at engaging international partners, a lunar-based architecture is most likely to emerge as the next focus of human space exploration. In addition, a lunar focus would provide practical opportunities for using private sector initiative, e.g., cargo delivery to the lunar surface. This could be done in a manner similar to International Space Station cargo delivery, but it would represent at least an order of magnitude greater addressable market even for an initial lunar base with the same number of crewmen as the Station.[15] In addition, new information from lunar robotic missions has strengthened scientific motivations to explore the Moon further. Russia has proposed an international lunar program with the United States and publicly maintained this position at international conferences. There are many geopolitical, scientific, exploration, commercial, and educational objectives that could be achieved at the Moon. To forego the opportunity for international collaboration to explore the Moon in favor of an asteroid mission, where there is little interest and no compelling objectives for a human mission, is a policy that is unsupported by technical or international realities.

Human space exploration is at a crucial transition point with the end of the Space Shuttle program and the lack of clear objectives beyond the International Space Station. At the same time, new space actors are present who lack the operational experience of major space projects with the United States. However, these actors have the potential to affect the sustainability, safety, and security of the space environment and thus impact U.S. interests in space. The seemingly separate threads of human, robotic, civil, commercial, and national security space activities are in fact deeply intertwined with each other, both politically and technically. The United States can best advance its national interests through a more integrated

---

15 Michael D. Griffin, "Enabling Complementary Commercial and Government Enterprises in Space," IAC-11.E3.4.6, paper presented to the 62rd International Astronautical Congress (Cape Town, South Africa, October 6, 2011).

strategic approach to its national security and civil space interests. International civil space cooperation, space commerce, and international space security discussions could be used to reinforce each other in ways that would advance U.S. interests in the sustainability and security of all space activities.

Advancing U.S. Geopolitical Interests in Space

The Greek historian, Thucydides, often considered the first "realist" in the theory of international relations, cited the primary motivations of a state as being the accumulation of wealth, power and prestige, in order to ensure self-preservation. In more modern terms, we might talk about advancing national interest through commerce, international security, and "soft power" or cultural influence. The United States is a major world power, the most powerful single country in the world, and would seem to have more than sufficient wealth, power, and prestige. However, the current state of relative U.S. decline in international influence in space, resulting from an inconstancy of purpose, is potentially dangerous. A decline in influence may be a minor consideration for a small power, but it can result in major geopolitical shifts for a major power.

The design of a geopolitically driven space program for the United States will be based on the unique needs and conditions of the United States. The interests of the United States are served by a stable international environment, a network of allies that trust and rely on U.S. security guarantees, global economic growth, and the acceptance and spread of U.S. values such as democracy, market economies, and tolerant, pluralistic cultures. Many countries share them and would embrace opportunities to pursue them mutually. In advancing those interests, a simple, unilateral approach is in fact unrealistic as the United States today is inherently a global power. The hallmark of U.S. foreign policy has always been its changing mixture of realism and idealism, what some might call its "Kissingerian" and "Wilsonian" aspects. In designing a U.S. portfolio of national space efforts, the needs of both "realpolitik" and idealism should be addressed.

Other than returning humans to the Moon as part of a broad, international effort, what else should the United States do to promote its geopolitical interests? How can it address other, overlapping challenges confronting national security and civil/commercial space activities?

Meeting Challenges in National Security Space

The first and foundational challenge is to strengthen U.S. space situational awareness (SSA) capabilities to cope with the increasing complexity of space operations. SSA is crucial to both national security and peacetime operations in space to avoid collisions among any nation's space object – not just those owned by the United States. Space debris from any source is a hazard to the ability of the United States to conduct space activities. SSA is also crucial in crisis and wartime in order to detect, characterize, and attribute hostile actions and enable appropriate responses across the full spectrum of conflict.

The United States needs to have sensors in multiple areas of the world to support SSA and increasingly needs to access information from other countries and the private sector. Given the criticality of SSA to national security space interests, it is unlikely that the United States would be willing to depend on an international organization for SSA data. Given the need to work with a range of other actors in managing space assets and responding to potential collisions, it is also unlikely that the United States will be able to operate in isolation from other spacefaring states. A possible solution to this dilemma is to take a hybrid approach with growing "circles of trust" in which SSA data and responses would be shared with others. Some information would be exclusive to the United States, some would be shared with close allies with whom close intelligence sharing arrangements already exist, some information would be shared on a reciprocal basis with other government and satellite operators, and some information would be made publicly available.

The second challenge is for the United States to develop and implement a national security space strategy for Asia. In particular, to build on existing alliance relationship such as with Japan, Australia, and South Korea to include dual-use space technologies. There is already extensive cooperation in the use of GPS, satellite-derived weather data, and satellite communications for military purposes. This cooperation needs to be extended to other space functions as well. In addition to SSA, data from space systems are needed to improve missile warning, ballistic missile defenses, and maritime domain awareness (e.g., monitoring of ships of all types).

The relative balance of power in the Asia-Pacific region is shifting as a result of the rise of China. The time when the United States might have been able to guarantee regional security largely on its own is closing. This does not mean the United States will retreat from Asia. Nor does it mean that China can be contained. Neither is feasible nor desirable for U.S. interests. Rather it means that the United States needs to create a stronger network of space-based informational sharing and utilization with and among our friends and allies in the region. The United States will remain the indispensible nation, but as the central "node" in a strong regional network rather than a hegemonic power.

China is not the Soviet Union. However strained relations may be from time to time, they are better than the continually hostile tenor of the Cold War period. The United States is economically and culturally engaged with China to a degree that it never was with the Soviet Union. As a global power, the United States needs predictability and order in international relations. As Robert Kagan said in the late 1990s, during the dot.com boom:

> "The people in Silicon Valley think it's a virtue not to think about history because everything for them is about the future, but their ignorance of history leads them to ignore that this explosion of commerce and trade rests on a secure international system, which rests on those who have the power and the desire to see that system preserved."[16]

Silicon Valley has increased its appreciation of international trade rules and foreign policies in more recent years. The global information infrastructure and the world economy depend on the uninterrupted functions of space systems of all types. This is particularly true for developing countries that cannot afford the installation of expensive terrestrial fiber optic networks or lack sufficient stability to protect large numbers of fixed towers for communications. As such, there are potential areas of common interest between developed and emerging economies in protecting and ensuring the sustainability of the space environments. This would be an opening for expanding adoption of norms of behavior as well as transparency and confidence building measures (TCBMs) that would make space a more explicit part of a secure international trading system. This would mean not just avoiding orbital collisions, but also reducing and mitigating radio frequency interference with space services. A satellite in space is useless if it cannot communicate with the ground – potentially as useless as if damaged by space debris.

---

16 As quoted by Thomas Friedman in "Foreign Affairs: Techno-Nothings" opinion article, *New York Times*, April 18, 1998.

In contributing to stability, TCBMs and accepted norms of space conduct can help reduce the chances of accidental conflict and provide cues to unusual activities. They cannot, however, substitute for the military capabilities necessary to deter potential adversaries. Deterrence in space is no different from deterrence on the land, seas or in the air: the focus is on understanding the thinking of an opponent. Ensuring adequate military capabilities requires understanding of how space systems fit into joint and combined arms campaigns, as well as understanding the views and values of potential adversaries. This in turn may well suggest steps toward greater international cooperation with friends and allies in order to enhance the potential effectiveness of U.S. military operations both in space and on Earth.

Meeting Challenges in Civil/Commercial Space

The current capability-driven approach to space exploration should be replaced by one that is driven by the geopolitical interests of the United States and its allies. As stated earlier, U.S. national space policy should be updated to more explicitly recognize international partners in a long-range vision of human space exploration. More specifically, U.S. human space exploration efforts for the next steps beyond low Earth orbit should be structured so the United States is at the center of a broadly international lunar architecture.

In particular, current language in the National Space Policy that directs NASA to send human to asteroid by or after 2025 and to orbit Mars by the mid-2030s should be deleted as it holds no interest for potential partners. Language from the NASA Authorization Act of 2010 could be adopted instead and thus bring White House and Congressional policy directions into closer alignment. Example text could be:

> NASA's human space flight and exploration efforts should enable the expansion of permanent human presence beyond low-Earth orbit and to do so, where practical, with international partners.

In terms of specific technical objectives, the NASA Authorization Act already has language that identifies several specific items:

> NASA should…determine if humans can live in an extended manner in space with decreasing reliance on Earth; identify means for meeting potential cataclysmic threats; explore the viability of and lay the foundation for sustainable economic activities in space; advance our knowledge of the universe; support United States national and economic security and the United States global competitive posture, and inspire young people in their educational pursuits.

Constraints on government budgets are such that private sector initiative, partnerships, and competition will be of increasing importance to many (but not all) space activities. In recognition of this fact, international discussions of space cooperation should also include measures to create greater stability, in both regulatory and policy arenas, in order to provide greater encouragement of private space activities. Legal support for the private utilization and exploitation of non-terrestrial materials and functional property rights should be part of incentives for space commerce and development. For example, one could imagine NASA purchasing privately provided water from lunar ice sources to support an international lunar research station. Other participating countries could do the same and thus provide an initial market for privately developed lunar resources.

In order to encourage states to align their space activities with ours, the United States could promote aerospace "free trade zones" with close allies. Just as in SSA "circles of trust," technology transfer regulations and labor movement rules could be eased in specific categories of commerce. There is precedent for this in defense cooperation between the United States and Canada and NORAD. The duty free movement of defense articles (e.g., electronics for the Distant Early Warning line) gradually expanded to include dual-use items and even automotive parts. This helped lay the foundation for what became the North American Free Trade Agreement. Clearly, there will be many countries for which this will not be practical in space technology, but creation of closer economic ties with important allies would help strengthen U.S. influence in space.

In general, U.S. geopolitical interests are advanced in concert with U.S. economic interests. The United States should encourage the growth and commercial competitiveness of U.S. space industries. NASA can and should take on diverse role in support of space commerce, e.g., through R&D, the reduction of technical risk, being a first or on-going customer for routine goods and services, and facilitating appropriate regulatory oversight by other federal agencies. NASA and the Department of Defense should not preclude or deter commercial space activities except for reasons of national security or public safety.

The United States is doing relatively well in space commerce related to information services, such as direct-to-home TV, GPS applications, satellite communications and remote sensing/GIS. On the other hand, the United States has largely been driven from international commercial launch services markets by Russia and Europe. As a result, the U.S. space launch industrial base is reliant on U.S. government purchases of launch services. Those services in turn cost more due to having few launches over which to amortize fixed costs. It would be very useful if U.S. firms could, without direct subsidies, regain international market share in space launch.

Space launch is fundamental to the United States being a spacefaring nation and the ability to place humans and cargo into space should be considered a strategic asset much like building nuclear submarines or high-performance aircraft. The existence of a market for private space transportation does not obviate the need for government-owned space transportation. Conversely, government space transportation should not compete with private providers (for example, through subsidized pricing). The United States requires an evolving mix of public and private space transportation options. While it is common wisdom to place the dividing line between the two at low Earth orbit (e.g., private launches exclusively to LEO and government systems beyond), other divisions could be considered. For example, there could be a criticality division depending on the mission. As mentioned earlier, private systems could be relied on to place cargo on the Moon, while government systems would be used to transport humans to the Moon.

The current U.S. policy of relying on Russia for crew access to Earth orbit while U.S. private sector providers develop crew transportation capabilities demonstrates a lack of seriousness toward human access to space as a strategic capability. It is certainly possible that private providers will be able to recreate a capacity that the U.S. developed prior to the Shuttle. However, the United States has no alternative but to accept any delays, failures, or cost growth. One can only conclude that under current policy, human access to space is desirable but not truly crucial to the nation. This is indicative of a civil space exploration effort lacking a strategic rationale. Should the United States adopt a more strategic approach, it should ensure that it has a sustainable mix of both public and space transportation options for assured access to space. A clear space transportation architecture supporting an international return to the Moon would enable optimizing that mix of capabilities.

NASA and the Department of Defense should cooperate on common approach to sustaining the U.S. space launch industrial base through block buys and use of common engines, subsystems, and components wherever possible.

Conclusion

There are many disconnects among U.S. space policies, programs, and budgets. Most seriously, there is a lack of strategic purpose to our human space exploration activities and a lack of coherence to our national security space strategy. Our civil, military, intelligence, and commercial space sectors operate independently of each other, despite confronting common global conditions abroad and reliance on the same industrial base at home.

Even in a tight fiscal environment, the United States can advance its national interests more effectively by taking a more integrated approach to its space capabilities and international cooperation. It can and should avoid unrealistic and dangerous hopes that other nations will naturally align their interests with ours in space. The current situation is one in which other nations are drifting away due to failures in leadership. It will take active leadership to repair the damage. In a geopolitical and strategic approach to space, political leadership is needed to project a positive vision of the future and not merely one of managing decline gracefully. Positive, inclusive leadership in space will attract others to us and benefit the United States and its allies in terms of increased security, commerce, and global influence.

Reductions in human space exploration, defense budgets, and our international commitments may seem necessary to redirect resources toward domestic needs. Such strategic withdrawals have occurred in other countries, such as the United Kingdom after the Second World War.[17] This is not a realistic option for the United States. The United Kingdom could rely on us to take over responsibilities for global stability. There is no one other than the United States, our allies, and the international order that has been so painfully sustained by American power, to protect our interests today. Hopeful talk about "space as a global commons" and "entanglement" and norms and codes as guarantors of security becomes dangerous when treated as substitutes for American power. As Thomas Friedman once observed on the role of global institutions and markets: "The hidden hand of the market will never work without a hidden fist."[18]

U.S. space activities should be in the service of national interests, and in particular, the interests identified by the first realist, Thucydides: wealth, power, and prestige. Without space, including a human space flight program that supports geopolitical goals, we will be assured of having none of these things. Without attractive and inclusive space leadership, we will not be able to shape the international environment for the space activities that our economy and security depend on. If we cannot ensure our international security and grow our economy, in the end we will not be able to sustain the social welfare benefits we have now.

Space activities do not fit within a single policy domain, department, or agency, but it is the very fact that they engage so many aspects of a nation's policymaking that make them so critical and beneficial to the nation. In shaping the international environment for space activities,

---

17 Mark Steyn, *After America* (Washington, DC: Regnery Publishing, 2012): 202. "After empire, Britain turned inward: Between 1951 and 1997 the proportion of government expenditure on defense fell from 24 percent to seven, while the proportion on health and welfare rose from 22 percent to 53. And that's before New Labour came along to widen the gap further. Those British numbers are a bald statement of reality: You can have Euro-sized entitlements or a global military, but not both."

18 Thomas Friedman, "A Manifesto for the Fast World," *New York Times Magazine,* March 28, 1999.

hard and soft power can be used to complement each other and build a more secure, stable, and prosperous, world in which our values are taken beyond the Earth.

Ensuring Space Leadership: A 21st Century R&D Investment Strategy

William B. Adkins

Throughout history, superior technology has been a decisive factor in nearly every human endeavor. Whether in military conflict, manufacturing, commerce, medicine, or any other field, those with a technological edge and the wherewithal to use it have most often prevailed. Advances in technology have transformed and improved the quality of life and propelled economic prosperity, often tipping the balance of power in the world. Since World War II, the United States has enjoyed an edge over the rest of the world economically, militarily, and scientifically -- largely due to investments in research and development (R&D), available resources and capital, and a free market system that rewards innovation and efficiency.

However, technology is fickle: the tables can turn quickly and without warning. For example, Microsoft was caught entirely off guard when the Internet browser Netscape burst onto the scene in the mid-1990s and forever changed how computers would be used. Some might say Microsoft has never quite recovered. What might be cutting-edge one decade could be passé the next. Big U.S. automakers, once dominant forces in the market, have largely ceded their lead to imports. Technological superiority is perishable, and staying ahead of the rapidly escalating pace of change will be a dominant challenge for the U.S. in the next 50 years. Increased emphasis and investment in R&D is essential and urgently needed.

R&D planning is important, but centralized planning is usually predicated on linear extensions of the current path and does not easily accommodate the serendipity and unpredictability inherent in R&D. Countless technological advancements originated as accidental discoveries where a researcher, scientist or inventor may have been trying to do one thing and found something else, often totally unrelated. Accidental discoveries include vulcanized rubber, penicillin (and many other drugs), numerous plastics and synthetic materials, radar, and the microwave oven, just to name a few.

According to science historian James Burke, who hosted the PBS series "Connections" on science and invention, hydrazine, a propellant commonly used on spacecraft today, started out as a fungicide in the 19th century to kill parasites in French vineyards.[19] So, somewhere along the line someone, perhaps Von Braun, had a flash of insight that it would make a good rocket propellant, and during World War II it was adapted to become a propellant. In the 1970s, rocket scientist Eckart Schmidt of Olin Aerospace discovered, though a chance mix up, that certain grades of hydrazine (Ultra-Pure hydrazine) would reduce demand on thruster heaters and significantly extend spacecraft thruster life. As a result of this serendipitous discovery, the propellant for the Voyager missions was changed to the Ultra-Pure hydrazine and the life of the Voyager missions was extended. Ultra-Pure hydrazine is now the only grade of hydrazine available in the U.S.[20]

Long-term success in advancing the frontier of technology requires an uncanny ability to anticipate what might be over the horizon and the

---

19 James Burke, Connections III, episode #4, "An Invisible Object", PBS TV series.
20 Eckart Schmidt, "History of Hydrazine Monopropellant," Presented at AIAA Pacific Northwest Section Young Professionals Technical Symposium, Seattle, WA, November 7, 2009.

agility to react to it, or ideally be the one driving it. Superior technology is not about the awesome complexity of a new "bleeding-edge" system – in fact complexity should be avoided – rather it is about faithfully applying Occam's razor: the simplest solution is usually the best. [21] Fundamental breakthroughs rely on ingenuity, flashes of insight, and cleverness in applying old ideas in novel ways. Sustaining a productive, innovative R&D effort over the long haul is as much an art as it is a science, requiring leaders adept at managing the unmanageable, willing to take risks, and creating sense out of chaos. To be successful, an R&D enterprise requires patience and an unwavering long-term commitment from policymakers, program managers, industry players, researchers, academia and the public. The payoff, though, is generous. Those with the best technology will succeed and prosper.

Nowhere is technological prowess and ingenuity more critical or more boldly displayed than the space program. Space activity requires an immense commitment of resources, unique facilities, and an extensive, diverse and highly-specialized supply chain — from large system integrators, to rocket propulsion suppliers, to specialty suppliers in nearly every discipline imaginable — and that is only the space segment. It also involves extremely sophisticated ground systems, state-of-the-art information technologies, massive data processing systems and millions of lines of software code. For a space mission to be successful, all of these things must operate properly together under highly stressing conditions and in extreme environments.

Space activity is among the most demanding multi-disciplinary technical endeavors and relies on a healthy pipeline of imaginative new ideas and innovative new technologies to meet the demands of increasingly complex missions. Of course, attracting and sustaining highly skilled

---

21 Occam's razor, accessed at http://en.wikipedia.org/wiki/Occam%27s_razor.

people, both in government and the private sector, with deep technical knowledge and expertise in managing large complex projects is at the core of maintaining a successful space program. Success also requires a culture that fosters and rewards creativity and innovation, and leadership that protects and nurtures promising ideas even when new ideas threaten the status quo.

Over the past 50 years, the U.S. space program has been a crown jewel of the nation. Space has gone from a technological novelty to an integral part of modern life, ingrained in nearly every facet of our lives. Space systems have become central to our national security, economic prosperity, and scientific advancement, both directly by the information gathered from space (weather, environmental monitoring, intelligence, missile warning, astronomy, planetary exploration, etc.) and indirectly by the services enabled by space (world-wide instantaneous communications, GPS-based navigation, etc.).

On the civil space side, NASA has been tremendously successful in accomplishing many astounding missions. Some of NASA's most remarkable accomplishments include the fundamental transformation of human understanding of physics of the universe from systems like Hubble and the Chandra X-ray observatory; the first probe to the outer planets (Pioneer); the first to leave the solar system (Voyager); the first to land on Mars (Viking); the first reusable launch system (Space Shuttle); largest international space project (International Space Station), and, of course, the first to land humans on the Moon with Apollo, a seminal moment in human history. Of course these are only a few highlights from an extensive and continuing legacy. Looking forward, NASA aims to answer fundamental human questions, such as: Has life existed elsewhere? What is dark energy? What is our destiny?

The national security community has a comparable list of its own. National security space systems have made a deeply profound impact on the nation's security and welfare. For example, U.S. space reconnaissance satellites helped win the Cold War; satellites showed that the 1950s U.S./Soviet "missile gap" did not exist and greater transparency provided through space systems helped provide stability and contributed to avoiding a major head-to-head clash. Today, space systems provide vital information and intelligence to support senior decision makers and policymakers in myriad ways and are an integral part of the nation's warfighting capabilities.

Yet, as important as space has become as a foundational capability, the research and technology base supporting it has become weak and fragile. The space technology base represents only a small sliver of the defense industrial base, which itself is only a small part of the nation's overall technology and manufacturing base. Over the past several decades, the U.S. industrial base has shifted emphasis away from manufacturing and toward a service and information economy. Specifically, the U.S. manufacturing base has declined as a share of the total economy from about 25 percent of the U.S. Gross Domestic Product (GDP) in 1970 to about 12 percent of GDP today.[22] Over the same period, the U.S. has lost more than 7 million manufacturing jobs.[23] While the share of the economy devoted to manufacturing has decreased, total manufacturing output

---

22 Mark J. Perry, ""Decline of Manufacturing" is Global Phenomenon: And Yet the World Is Much Better Off Because of It," American Enterprise Institute, April 29, 2011, accessed at http://www.aei-ideas.org/2011/04/decline-of-manufacturing-is-global-phenomenon-and-yet-the-world-is-much-better-off-because-of-it-2/.
23 Mark J. Perry, "The Demise of America's Manufacturing Sector Has Been Greatly Exaggerated," American Enterprise Institute, January 20, 2011, accessed at http://www.aei-ideas.org/2011/01/the-demise-of-america's-manufacturing-sector-has-been-greatly-exaggerated/.

remains very high because of significant improvements in efficiency/productivity resulting from automation and advancements in manufacturing technologies — which is great for mass-produced items. The changing nature of the manufacturing base may benefit some sectors where there is a large consumer base, but, potentially, it leaves highly specialized niche areas vulnerable, such as shipbuilding, military aircraft, and satellites. Without dedicated, targeted R&D and technology investments, these niche areas will likely stagnate.

In 2011, former DOD acquisition official Jacques Gansler shared with Congress his concerns about the current state of the defense industrial base: "… the U.S. industrial base that supports it [defense] has simply been consolidated from around 50 major suppliers to a half a dozen." A 2008 Defense Science Board task force that I [Gansler] chaired concluded: 'The Nation currently has a consolidated 20th century industry, not the required and transformed 21st century national security industrial base that it will need in the future.'"[24]

Space programs, and particularly research and technology investment related to space, continue to be dominated by government-funded programs. According to the Space Foundation, the U.S. government spends about $5.3 billion per year on satellite manufacturing, and the commercial sector spends about $5.6 billion per year.[25] While, at first blush, these look comparable, government and space programs tend to be very different. Government space programs perform a very wide range of mission types, including military communications, GPS, weather, intelligence, remote sensing, planetary exploration, astronomy, and human spaceflight. Government space agencies tend to build unique, one-of-a-kind, spacecraft that stress the limits of technology and capabilities. In

---

24 Jacques Gansler, Testimony before the Senate Armed Services Committee, Hearing "The Health and Status of the Defense Industrial Base and its Science and Technology-related Elements," May 3, 2011.
25 The Space Foundation, *The Space Report* (Washington, DC: The Space Foundation, 2009): 16.

short, they drive technology. The commercial sector, on the other hand, generally focuses on only a few mission types: primarily satellite communications and remote sensing, and tends to be more production oriented; commercial communications satellites are essentially a commodity, whereas the government spacecraft tend to be a boutique development that demand and drive technology.

As such, government space programs have been particularly vulnerable to vacillating policies, ever-changing strategies, and erratic funding. Retired aerospace executive Tom Young testified before Congress about the impact of ever-changing policies and programs, stating, "Resources in terms of money and human talent that have been wasted on cancelled projects and aborted strategy is a national embarrassment."[26]

Numerous studies over the last several years are filled with admonitions about the current state of the U.S. space industrial base, the growing impact of an aging workforce moving ever closer to retirement, a technology base weakened by years of volatility from troubled programs, and anemic investment in research and development. In its 2012 review of NASA's technology program, the National Research Council (NRC) found that, "Success in executing future NASA space missions will depend on advanced technology developments that should already be underway. It has been years since NASA has had a vigorous, broad-based program in advanced space technology development, and NASA's technology base is largely depleted."[27]

---

26 A. Thomas Young, testimony before the House Space Subcommittee, Hearing "A review of Space Leadership Preservation," February 27, 2013.
27 National Research Council, *NASA Space Technology Roadmaps and Priorities: Restoring NASA's Technological Edge and Paving the Way for a New Era* (Washington, DC: National Academies Press, 2012): 1.

A 2009 NRC study on civil space echoed similar concerns, "The United States is now living on the innovation funded in the past and has an obligation to replenish this foundational element."[28]

Studies of the space industrial base by the DOD have made similar observations about the fragility of the industrial base, the aging workforce, and the damaging impact resulting from excessive program volatility: starts and stops, schedule delays, and cost overruns. A 2008 DOD study found that, "Successful programs are those that have realistic cost and schedule expectations, are well understood, have stable budgets, experienced and stable staffs, and have a spiral development acquisition strategy."[29]

The pattern that led to the depletion of the technology base seems clear. In short, fixing near-term problems consistently trumped the investments for tomorrow. Success-oriented planning needed to sell a major program led to over-promising capabilities and under-estimating technical, cost, and schedule risks. Once the underlying problems come into clear view, the inevitable pressure to get the program back on track ultimately robs the less flashy, long-term R&D and technology investment. It is shortsighted, but when managers are faced with the choice, it is also hard to blame them for making the difficult decision to keep the development program on track. Occasionally, the decision has been made to cancel a high-profile program, e.g., NPOESS, TSAT, and X-33. But the persistence of the pattern of bailing out troubled programs also suggests it has become a de-facto strategy. Moreover, the de-facto strategy creates a perverse incentive that can end up rewarding poor management. Over

---

28 National Research Council, *America's Future in Space: Aligning the Civil Space Program with National Needs*,(Washington, DC: National Academies Press, 2009): 57.
29 Steven R. Miller and John Thurman, *2008 National Security Space Industrial Base Study*, (Washington, DC: U.S. Department of Defense, 2008): 1.

time this pattern has left the technology base barren and nearly depleted, as confirmed by the aforementioned studies.

Instead of a de-facto strategy that uses R&D and technology accounts as a bill-payer, a clear and deliberate strategy is urgently needed to revitalize and strengthen the space program as a whole—from R&D through development and operations. It is not just a NASA or DOD problem, it is a national problem, and it will be a large factor in determining the nation's future in space.

Revitalizing R&D

So, starting from a clean sheet of paper and assuming that budgets remain tight for the foreseeable future, what policies and programs might be put in place to revitalize space R&D and technology development, improve program outcomes, and ensure U.S. space leadership? How should it be organized and funded to most efficiently develop innovative and useful technologies? How might priorities be set to determine what technologies to pursue? Can it be done in today's fiscally constrained environment? What might have to be given up?

Creating the innovative space technology engine the country needs will require far more than throwing a bunch of money at a technology program and hoping for the best. It will require a fundamental restructuring from the foundations in basic and applied research, to technology development, to structuring missions and flight programs in a manner that supports rapid technological advancement.

Most space R&D and technology program plans are organized according to the maturity level of the technology using a scale called the Technology Readiness Level (TRL) scale. The TRL scale goes from TRL-1 (least mature) to TRL-9 (most mature).[30] A technology program would typically start at the TRL-1 where basic principles are understood, and progress up the scale as the concept is matured, refined, and tested. But what happens when the system has run out of ideas that feed into TRL-1? Certainly, resourceful technologists will leverage new discoveries from other disciplines outside of space to build innovative new ideas, as they should. However, the unique needs of operating in space (zero-gravity, high radiation, etc.) may limit possibilities to adaptations of non-space applications. Space is a unique arena that requires a dedicated R&D and technology effort to produce relevant discoveries of new phenomena or scientific breakthroughs specifically with space applications in mind.

On the other end of the TRL spectrum, many technology projects that succeed through the early TRL stages with proof-of-concept and laboratory demonstrations never seem to make the transition to flight and operation, because proving a system in space involves a significant increase in cost. Also, spacecraft engineers and mission scientists tend to be reluctant to take on added risk by using their mission as a test bed for new technology. So, if the system feeding it (research) is not producing breakthroughs, and if the system needed to mature technologies (flight opportunities) are not available, the technology program risks creating a lot of "gee-whiz" hardware that struggles to make it out of the lab and into space. Consequently, new technologies with considerable potential remain on the proverbial shelf.

Removing or reducing impediments for innovative ideas to progress from initial principle through to flight will require a significant

---

30 Technology Readiness Level (TRL) is a measure used to assess the maturity of evolving technologies (devices, materials, components, software, work processes, etc.) during its development and in some cases during early operations. The TRL scale goes from 1 to 9, with 1 being the lowest level of technology maturity where basic concepts are understood. The scale progresses up to TRL-9 where the technology is proven on orbit through flight demonstration and ready for operation, http://en.wikipedia.org/wiki/Technology_readiness_level.

re-structuring of the entire spectrum of activities from research to flight operations. To accomplish this, research and technology advancement must become as important as performing flight missions. Activities might then be organized around the following three thrusts:

- Basic and Applied Research. Significantly expand the scope and scale of basic and applied research in fields relevant to applications in space technology, particularly high risk/high payoff areas done in-house by the government.

- Technology Development and Demonstration. Increase investment and priority for developing new systems, subsystems, and components. Provide strong incentives to bridge the gap between laboratory demonstrations and first flight, which technologists and engineers often call the "valley of death."

- Flight Missions and Operations. Tailor flight mission tempo, cost, and complexity to encourage the use of new technologies and concepts.

To succeed, a strong organization with significant technical and financial resources needs to take the lead. NASA, if appropriately vectored, may be the best organization to lead this enormous national task. Some could argue that the Department of Defense (to include the Intelligence Community) is a bigger player, with deeper pockets and better suited in this role, but space R&D would risk getting drowned out by other pressing operational priorities. While DOD is certainly a major force in the space arena, no organization within the national security community has the depth of people and facilities to match what NASA can offer in terms of a space R&D capability. Furthermore, security rules and regulations and a nearly impenetrable bureaucracy make it very difficult for DOD to reach beyond its traditional industrial base for innovation, with the possible exceptions of the Defense Advanced Research Projects Agency (DARPA) and some of the defense research laboratories. Those DOD organizations can and should play an important supporting role in revitalizing space R&D activities, but they are not well equipped to take the lead.

The private sector also has an important supporting role, especially in reducing risk as technologies mature from lab curiosities to flight demonstration and on to operational use and production. Incentives are critical for turning innovations and demonstrations to operational use, and eventually into production items. In general though, the private sector no longer takes the kind of risks needed in basic and applied research and technology development. Much of the private investment in space builds on existing technologies and concepts and focuses on providing a service to a market, rather than fundamentally new technologies.

As the lead agency, NASA would need to restructure and reprioritize many of its activities, though NASA's science and exploration programs and goals should not to be forgotten, either -- they are worthy endeavors in their own right. However, the approach to implementing science and exploration goals and priorities, such as those found in NRC decadal surveys or plans for human spaceflight, would need to be significantly restructured to increase the frequency of opportunities for technology insertion and to accelerate the rate of advancement. Under this plan, NASA's most important national contribution over the next decade would be as a technology engine for the national space enterprise as a whole. In fact, some could argue that, in the aftermath of Sputnik, technology advancement to catch up with the Soviets was a major factor in forming NASA in the first place. As in 1958, NASA may again be the appropriate place to build the nation's future in space.

Cultural Transformations At NASA

The first step is to transform the culture to value technological research and development and technical excellence as an equal priority to major flight missions. Moving the needle on culture is notoriously difficult, but once properly aligned, organizational culture is a powerful driving force. Culture determines values, guides behaviors, drives priorities, budgets, and program decisions, and ultimately sets the direction of any agency or organization.

NASA's early culture stressed technical excellence. In 1958, NASA was formed through the amalgamation of several world-class research institutions, specifically the National Advisory Committee on Aeronautics (NACA), the space-related activities of the Naval Research Laboratory, the Army Ballistic Missile Agency (ABMA), and portions of the Air Force. Each of these organizations was deeply involved in research and development built around a culture of technical excellence. Over a period of more than 40 years starting in 1915, NACA made revolutionary advancements in aeronautics that have had a lasting impact on U.S. capabilities for military and commercial aircraft. NACA won five Collier trophies (1929: low drag engine cowling, 1946: thermal ice-prevention for aircraft, 1948: Chuck Yeager's X-1 supersonic flight, 1951: supersonic wind tunnel, 1953: transonic "area rule," a principle critical for designing supersonic aircraft). Today, the U.S. has the most advanced military aircraft and leads the world in civilian aircraft because of the many advances made by NACA and NASA. ABMA, under the guidance of Dr. Werner Von Braun, brought decades of hands-on experience and the best rocket team in the world to NASA, leading, of course, to the Saturn and Space Shuttle programs.

NASA's extraordinarily rapid advancement in its first few decades was a direct result of the strength of the technical foundation of these groups. The current challenges facing the space program and the fragility of the technical base, in government and industry, are stark reminders of the perishability of technical expertise and capabilities. In many ways, reinvigorating NASA's technical capabilities is a return to fundamentals. Restoring the goal of technical excellence as a core mission, as was the case under ABMA and NACA, is a crucial first step.

It is not that the problem of the dwindling technology base has gone unnoticed. Over the past 20 years or more, various proposals and initiatives have been advanced to increase the priority of R&D and technology development at NASA. Few have gained much traction, with plans often scuttled over disagreements in approaches to programs and policies or because funds were siphoned off to address pressing cost issues on programs in the throes of development.

Recent examples include the initial iteration of the Vision For Space Exploration, when then NASA Administrator Sean O'Keefe and Exploration head Craig Steidle proposed a broad-based technology plan to invest in new ideas in nearly every aspect of spaceflight. Technologies and concepts would be selected as they matured and eventually incorporated into mission architectures. The underlying philosophy was to cast a broad net and winnow the group down to the most promising technological ideas. It was a very measured approach. However, it was critiqued as too slow and open-ended and was eventually replaced with a specific point design, Constellation, built primarily on existing technologies particularly from the Space Shuttle program. The Constellation approach was attractive because it appeared likely to fly sooner by building on heritage technology. After a few years and amid growing concerns over cost, the Obama Administration proposed replacing the Constellation program with a technology program built around "game changing" technologies. Congress largely rejected much of this approach in favor of a more mission-focused approach similar to Constellation with a much-scaled back technology initiative.

In February 2013, NASA announced the creation of a new Space Technology Directorate, a clear acknowledgement of the vital importance of technology and innovation to NASA's future. In announcing the new directorate, NASA Administrator Bolden said, "A top priority of NASA is to invest in cross-cutting, transformational technologies. We focus on collaboration with industry and academia that advances our nation's space exploration and science goals while maintaining America's competitive edge in the new innovation economy." It is certainly a move in the right direction, but it is too early to tell if it will have a lasting impact.

So, assuming — perhaps by waving a magic wand — that the culture is in fact transformed, and space R&D and technology development are fully embraced as top priorities, the new technology-driven space program, the "new NASA," might be organized around three thrusts: (1) Basic and Applied Research; (2) Technology Development and Demonstration; and (3) Flight Missions and Operations.

Basic and Applied Research—Built on NACA and DARPA Models

Basic and applied research are the most fundamental types of research and form the bedrock of long-term success. The Office of Management and Budget's (OMB) government-wide definition for basic research is: "systematic study directed toward fuller knowledge or understanding of the fundamental aspects of phenomena and of observable facts without specific applications towards processes or products in mind. Basic research, however, may include activities with broad applications in mind."[31] Applied research is defined as: "systematic study to gain knowledge or understanding necessary to determine the means by which a

---

31 OMB Circular No. A-11, 1998.

recognized and specific need may be met."[32] To some, research may seem like a waste of time and money, open-ended, overly driven by individual curiosity, or even boring. Research is not nearly as exciting as a flight mission, but it is the most elementary ingredient for any technically complex activity. A healthy and re-energized research program will sow new seed corn and is the best first step in revitalizing the crop of innovative ideas for the years ahead.

Some may also argue that basic research is the responsibility of other agencies, such as the National Science Foundation, DARPA, the National Labs, or DOD Labs, and that NASA is a "mission-driven" agency, not a research agency. The fact is that those agencies have broader portfolios than space and have more limited resources in terms of people and facilities to apply to the task. The current state of the space industrial base is evidence — or even proof — that the current research apparatus is not working to keep the U.S. at the level it needs. A new approach must be taken. As stated earlier, NASA was built on the foundations laid by strong research organizations, like NACA. Many of the advances made in the first few decades of the space program were built on those foundations, which are now weak and in need of reinforcement.

The proposed plan would create a hybrid research and development organization that builds on the depth and continuity of the NACA model and the flexibility and agility of the DARPA model.

Space historian Roger Launius described the NACA model in an April 11, 2012 *Space News* editorial.[33] In short, the NACA model involved soliciting research ideas from all sources, including the military and other government agencies, private industry, and the NACA staff. Requests from the military and other government agencies were generally auto-

---
32 OMB Circular No. A-11, 1998
33 Roger D. Launius, "The NACA Model for Technology Transfer," *Space News*, April 11, 2012.

matically accepted and NACA would carry them out, mostly using funds appropriated to NACA. Other proposals were reviewed for technical merit and those selected were accomplished "in-house" by government scientists and engineers with the final product being the results of the investigation, test, or study. NACA had major facilities at what are now NASA Langley, Glenn, Ames, and Edwards research centers and over 7,500 employees before it was absorbed into NASA in 1958. Over four decades, NACA scientists and engineers produced more than 16,000 research reports, notes, bulletins and memoranda. According to Launius, the reports and papers "became famous for their thoroughness and accuracy, and served as the rock upon which NACA built its reputation as one of the best aeronautical institutions in the world."[34] NACA was a large organization with deep technical talent, decades of experience, and world-class facilities.

DARPA operates quite differently than NACA, yet it also has an impressive record of innovations over its 50-year history, e.g., ARPANET/Internet, Unmanned Aerial Vehicles (UAVs), stealth aircraft. DARPA is known for its "out-of-the-box" thinking and willingness to pursue high-risk/high-payoff ideas. DARPA is a small organization of highly trained technical experts with a total staff of about 250, roughly 150 of who are technical managers on rotations of four to six years from other institutions and agencies. DARPA is a very flat organization with minimal bureaucracy. It has no facilities or laboratories and projects are done "out of house" under contract to industry or by government research laboratories. DARPA was created in 1958 to "prevent strategic surprise from negatively impacting U.S. national security and create strategic surprise for U.S. adversaries by maintaining the technological superiority of the U.S. military."[35] Its activities range from basic research up to full-scale technology demonstrations. DARPA's annual budget is about $3 billion,

---

34 Launius, "The NACA Model."
35 DARPA website, http://www.darpa.mil/our_work/.

though it is spread across a very broad range of areas, including materials, electronics, information technologies, biotech, sensors, platforms and space. The FY13 budget request for DARPA space programs was about $160 million.

Both the NACA and DARPA models have attractive features. While some remnants of the NACA model remain at NASA, that model of technology development no longer dominates the agency's culture, operations, or planning. There is a range of reasons for this, but the consequences for the industrial base are unavoidable and unacceptable. A concerted effort to revitalize it would provide a strong base of core talent, a brain trust of expertise that over time would become a national asset and resource available for all space programs. A DARPA-like organization with its iconoclastic culture and regular infusion of fresh blood could provide an avenue for unorthodox thinking and revolutionary advancements.

Can these two models be combined under one roof at NASA? Absolutely, and doing it in a manner that preserves the strengths of each model and creates constructive tension between the two could magnify the benefits even further. If so, how might it work?

A NACA-like organization, with its large workforce and major facilities, would seem to fit well as a stand-alone Directorate. The Space Technology Directorate proposed earlier this year by NASA could be morphed into the NACA model. However, far greater emphasis would be placed on basic research than is currently the case. Like NACA, a key product of the activity would be conducting research and development and providing results to its sponsors in technical reports and studies and making it available to the larger space community. The research priorities of the "NACA-Directorate" should be determined by the researchers themselves, who would be allowed to follow their instincts and would be trusted to carry out their work with minimal oversight. They should be encouraged to take risks; management should judge the results of the NACA Directorate as a whole not as individual projects, as only 1 in 100 ideas may fully succeed, but those rare successes are likely to be transformational.

A few adjustments to the traditional NACA model might be considered in applying it today. Allowing other space agencies, such as the Air Force, the National Reconnaissance Office, or the National Oceanic and Atmospheric Administration, to directly task the "NACA Directorate" with research projects would be helpful. The main difference would be requiring outside agencies to share in the cost, such as a minimum of 50/50 cost share. While this would force NASA/NACA to budget for work for others -- and could raise uncomfortable tensions and potential clashes between differing cultures and mission objectives -- the current approach, with agencies tending to operate solely within their own "sandbox," risks missed opportunities for significant technology advancement that support the space program as a whole. Encouraging collaborative efforts with cost sharing would create an incentive for space agencies to work more closely and effectively together in areas of common interest. It also could help to more fully utilize what might otherwise be underutilized facilities. Also, by aggregating more work, it would help build up and maintain a critical mass of expertise, as well as serve as an excellent mechanism to share information and results to help foster technical excellence. The private sector could task NASA/NACA through current mechanisms, such as Cooperative Research and Development Agreements (CRADAs). The NACA-like Directorate would, of course, also have all the customary traditional contracting tools, as well as non-traditional mechanisms, e.g., Space Act Agreements, to work with the private sector

The magic ingredient is the quality of the research staff. To be successful, it is absolutely necessary to attract and retain a strong in-house technical staff with ample hands-on experience. A few good technical leaders are essential. A 1980 study noted the importance of having a few exceptionally talented individuals:

> The presence of a few individuals of exceptional talent has, to a very large degree, been responsible for the success (and even the existence) of outstanding research and technology development organizations. It is not the function of the laboratory to be 'representative,' to be a cross section of the population, but to nurture exceptional talent wherever it may be found. A technology development organization that cannot tolerate and nurture a few eccentrics is halfway toward rigor mortis. In many Federal laboratories, the best fundamental research projects are almost always built around one outstanding individual. This person has demonstrated that he is capable of performing basic research of high quality, and he is accordingly granted the freedom to pursue his interests.[36]

Fostering and developing such an environment, with great technical leaders, would be critical for the NACA-like Directorate. Attracting great technical leaders into government can be challenging, but in 2004 Congress provided NASA with a variety of tools and authorities to help attract and retain top-notch technical talent, including enhanced recruitment and retention bonuses and special exceptions to federal pay caps. NASA should make full use of these authorities. Furthermore, NASA has not had the authority in recent years to shape and "right-size" its workforce through reductions in force. Attracting and retaining the best technical talent into government is central to the NACA-like Directorate's success.

---

36 Hans Mark and Arnold Levine, *The Management of Research Institutions: A Look at Government Laboratories* (Washington, DC: NASA, 1984): 156.

In contrast, a DARPA-like organization would seem to fit best as an activity separate from the rest of the agency, but reporting directly to the NASA Administrator. NASA could essentially take the DARPA model in its entirety and implement it, recreating the latter's tolerance of risk (and failure), flexibility, and high degree of autonomy. To be clear, this organization is not to be a staff function, but would have contracting authority and responsibility to manage and run projects in the same manner as DARPA does currently. A DARPA-like organization within NASA should be independent of any responsibility for laboratories, facilities, and centers. The main difference would be that the NASA version would focus on areas of potential interest to space and space-related problems and areas.

Some could argue that having both a NACA-like Directorate and a DARPA-like office within separate reporting chains would be an inefficient organizational approach and could create conflicts over missions, goals, ideas, people and turf. Yet creating some constructive tension and competition for ideas is precisely the point. Providing the senior leadership at NASA with alternative avenues to conduct business and alternative sources for ideas and opinions might help promote a deeper dialog over the choices of direction.

Technology Development and Demonstration

While the basic and applied research activities are intended to be self-organizing and self-directing, technology development and demonstration activities would be planned through a more traditional road-mapping process, such as the technology roadmap completed by the NRC in 2012 and the follow-on strategy and plans released by NASA.

A key challenge for technology development is traversing the "valley of death" and getting new technologies implemented on projects. Often the Principal Investigator or Program Manager is doing everything possible to reduce risk. While a new technology may show great promise for significant performance benefits, it may be overshadowed by the perception of increased risk and potential for failure. Technology projects have an uncanny knack for delays and technical glitches that make them unattractive for flight on a mission, especially if it has a narrow launch window that might not return for years, as planetary missions sometimes do.

NASA has tried in recent solicitations to facilitate the infusion of new technology by including cost incentives. Specifically, last year's Discovery-class mission solicitation included a menu of new technologies for potential bidders to consider including in their design, such as the NASA's Evolutionary Xenon Thruster (NEXT) and the Advanced Material Bipropellant Rocket (AMBR). Discovery missions are cost capped at $425 million (not including launch). As an incentive to incorporate new and unproven technologies, NASA formally offered in the solicitation to increase the cost cap for the project by $19 million if the proposing team included the NEXT thruster and $5 million if it included the AMBR technology. While this does not guarantee that the new technologies will be in the proposal selected, it seems like a good common-sense way to help bridge the gap and move technologies out of the lab and into operation.

Such an approach should be used more broadly and could be extended beyond NASA to include other space agencies' technology programs. Stepping up activities to find flight opportunities for new technologies could be of great benefit. The burgeoning world of hosted payloads and rideshare opportunities also provide a viable avenue for technologists to seek out flight opportunities.

Flight Missions

The area most impacted by refocusing on research and technology development is flight missions. In this discussion, flight missions encompass missions in NASA's Science Mission Directorate and missions in human spaceflight. This analysis is based on a "clean sheet" of paper, so it focuses on the attributes of the desired end-state, and is not intended to prescribe how to make the transition from the current portfolio.

As stated earlier, it is not enough to simply bolster efforts in research and technology development to revitalize the system. A robust program of regular flight opportunities is a critical final leg in the technology development and maturation cycle. While a dedicated flight technology program, similar to the Air Force's Space Technology Program (STP), should be undertaken with at least one such launch each year, a re-architecture of the approach and philosophy for flight missions is needed to ensure lasting impact and to provide stability.

The trend in the last few decades, across civilian and national security space, has been toward a fewer number of highly capable but very expensive space systems. Indeed, these systems are spectacular and offer tremendous capabilities if/when they are ultimately fielded. They are exquisite. They are technological masterpieces. But they come at a price.

The danger with very large, complex development projects is that they create a vicious cycle of low quantity and high cost which magnifies the reluctance to take risk, squeezes out opportunities for technology infusion, and reduces the frequency of mission opportunities. This cycle can spiral out of control as cost, complexity, schedule, and oversight compound to make a program unaffordable. A portfolio dominated by only a few large programs can force managers to delay investment in technology, delay smaller missions, and slow the mission development and launch tempo to a near-standstill. The reduced demand affects the whole supply chain and exacerbates weaknesses in the space industrial base, especially among specialty suppliers, e.g., radiation-tolerant parts, power systems, propulsion, or avionics.

Long, drawn-out program schedules also create challenges in maintaining continuity in program leadership and staffing, both in government and industry. Engineers and scientists fresh out of school may think twice about getting into a space career if they think they will spend an entire decade on just one program.

Mega projects have a tendency to gain a life of their own and grow beyond anyone's control. According to Donald Schon, an expert in organizational behavior:

> Large-scale developments of the kind undertaken by super-corporations or the military may proceed for months or years beyond the point where they should have stopped; they continue because of massive commitments to errors too frightening to reveal . . . In these cases, the personal commitment of the people involved in the development, the apparent logic of investment, and the fear of admitting failure, all combine to keep the project in motion until it fails of its own weight."[37]

These programs might be known today as "too big to fail and too big to bail." Such programs often leave a trail of destruction, with lost decades of work and unspeakable sums of money down the drain. Both the national security and the civilian space communities have had failed programs that fit Schon's description. In the end, a portfolio dominated by large projects with long development times risks stagnation and driving itself out of business.

Perhaps the most important factor in ensuring a strong and vibrant space program is flight rate tempo. A high flight rate tempo brings with

---

[37] Donald Schon, "The Fear of Innovation," *International Science and Technology 14* (November): 70-78.

it many advantages, including more frequent opportunities to infuse new technologies and spreading risks over a larger number of programs. Given tight budgets, a high tempo also implies smaller, less capable, missions on shorter, more manageable timelines and perhaps smaller incremental steps. Historically, periods of high flight rate tempo have coincided with rapid advancement, because there are more opportunities to try new ideas and concepts.

The decade of the 1960s was a period of extremely rapid technological advancement for the space program. Building off the foundations of research-oriented organizations like NACA and others, the space program advanced very swiftly. In 1960, space was still new and failure rates were very high. The tolerance for failure was also high and allowed engineers to learn rapidly from those failures. For example, the first 12 flights of the Corona satellite surveillance system failed before the first success. Eventually, Corona went on to make innumerable critical contributions to national security.

During the 1960s, U.S. technical proficiency in space grew exponentially. By the end of the decade, the U.S. had landed on the moon — a mission nearly inconceivable in 1960. Mercury, Gemini, and Apollo missions are among the most notable successes of the space program and made significant contributions to the technological foundation of spaceflight. These missions are only part of the story. Human spaceflight missions accounted for less than 30 of the more than 500 missions conducted by the U.S. in the 1960s. In fact, more than one-third of all U.S. launches to date occurred in the 1960s. It was a very high tempo. The launch tempo in the 1960s averaged about one launch per week and many launches had multiple satellites on-board. This is nearly three times the launch pace seen during the first decade of the 2000s. While satellites today are much more reliable, this differential begs the question of what the proper pace should be to strengthen the industrial base, reduce individual mission costs, and provide plentiful opportunities to infuse new technologies. The current flight mission pace coupled with anemic investment in R&D is an almost certain death spiral.

The Air Force seems to have recognized that its current architecture, predicated on a small number of highly capable systems, is leading to a weak and fragile industrial base, skyrocketing costs, and potential vulnerabilities to attack. In response, the Air Force is seriously examining "disaggregation." Disaggregation is aimed at splitting systems up into smaller, more focused missions. For example, instead of flying several weather sensors (some of which may not have been particularly compatible) on one large and fairly complex spacecraft, disaggregation would break the sensors up into smaller more focused missions, including free-flyers and hosted payloads on commercial spacecraft. While there may be some reduction in capabilities, the loss may be worth it. Simplifying the individual missions and spreading the risks over a larger number of platforms would allow for a more resilient overall architecture in terms of industrial base and cost stability, more frequent technology infusion, survivability, and scalability. It is not clear how far the Air Force will go. .

NASA went down a somewhat similar route in the 1990s when NASA Administrator Dan Goldin implemented the "Faster, Better, Cheaper" (FBC) initiative. While the mere mention of FBC often causes eyes to roll, there are some who believe the conventional wisdom is wrong and that a close examination of NASA's FBC experience shows it was a success -- and perhaps a model worth revisiting. For instance, Dan Ward, an Air Force acquisition expert, points out that the first 9 out of 10 missions under FBC were successes and provide proof that challenging missions can be accomplished in short timelines for moderate amounts of funding.[38] Ward also examined the six failures that occurred under FBC, which were, according to the failure report, "attributed to poor communications and mistakes in engineering and management." He contends that "such failures are arguably avoidable, but they are neither unique nor ubiquitous to the FBC method."[39] While the phrase "faster, better, cheaper" may not come back in vogue, the core tenets and philosophy should be revisited. As Ward points out, implementing FBC requires "an understanding that these approaches are not methods or processes, but rather something akin to a worldview. They are sociological and cultural — not procedural — approaches."[40]

Since the Apollo era, "failure is not an option" has been a guiding mantra. This mantra has been taken too far. Over the years, the culture has morphed from one of prudent risk management to risk aversion, and ultimately stagnation and near paralysis. Taking risks, and even experiencing harsh failures, are a vital aspect of a vibrant, healthy, and advancing organization. Effective technical and organizational leaders require the scar tissue earned via the hard-won lessons of failure.

With the lessons from the 1960s and FBC in mind, how should flight missions be approached in the future? Re-architecting NASA's flight missions should start with re-scoping expectations in the NRC's decadal survey process. The NRC's decadal surveys provide the science community's prioritized list of missions by scientific discipline. Developing consensus is difficult, but once achieved, the list serves as the foundation for plans, programs, and investments. The surveys have done an outstanding job listing the priorities, but they have missed the mark significantly

---

38 Dan Ward, "Faster, Better, Cheaper Revisited", *Defense AT&L*, March-April 2010, accessed at http://www.dau.mil/pubscats/ATL%20Docs/Mar-Apr10/ward_mar-apr10.pdf.
39 Ward, "Faster, Better, Cheaper," 51.
40 Ward, "Faster, Better, Cheaper," 52.

regarding estimating costs and time required to implement programs. For example, the James Webb Space Telescope (JWST) was significantly underestimated for cost, complexity, and schedule. Some might wonder whether the decadal survey would have ranked JWST so highly if the panel knew at the beginning that it would take nearly 30 years and $8 billion to complete.

A re-scoped NRC decadal survey, aimed at both scientific advancement and ensuring a healthy technology base for the future, might consider eliminating flagship missions from consideration and providing priorities organized by cost caps, such as small ($200 million), medium ($500 million), and large ($1 billion). While some may howl at prohibiting flagship missions and forcing science in a cost box, continuing down the path of the current de-facto strategy ultimately leads to a barren space program limited by 20th-century technologies and ideas.

The objective would be to increase the mission tempo by a factor of two, thus increasing the number and frequency of opportunities for mission scientists, engineers, and managers, balance mission type/destination, increase demand from the supply base to provide stability, and significantly increase opportunities to insert new technology. Also, a portfolio with a larger number of missions spreads the technical risk and makes it more palatable to make difficult decisions to re-scope or cancel missions that run into trouble. Ideally, program managers and scientists would begin to regain more control over their programs, allowing them to manage more effectively because smaller programs do not typically demand that the agency stake its reputation on them.

As noted earlier, a flight technology demonstration, similar to the Air Force's STP program, should be flown at least once a year to provide a dedicated mission for new technologies. These missions should be conceived, managed, developed, integrated, tested, and operated in-house by the "NACA Directorate." Establishing the "NACA Directorate" as a cradle-to-grave "skunkworks" organization with technical personnel involved across the spectrum of activities -- from research to hardware and system development -- is an essential aspect of ensuring that the government has a "smart buyer" and a training ground for promising managers, engineers, and scientists. Ideally, lead experts would have something brewing on each front to help keep them intellectually active and engaged. Moreover, such an approach would promote cross-pollination among multiple engineering and technical disciplines and fields.

Conclusion

The U.S. program still leads the world in space. However, U.S. space leadership should not be measured simply by individual program goals and accomplishments, i.e., landing on the moon or a planet, or building a big telescope. If leadership is measured only by those accomplishments, the nation will never evolve past the unaffordable, exquisite program, and it will fall deeper into a death spiral.

More fundamentally, leading the world in space is about possessing superior capabilities and providing the nation with the capacity to do whatever it wants, whenever it wants to do it, in space. Such a capacity depends on developing, maintaining, and sustaining a robust industrial base, attracting and retaining the best and brightest, and building trusted partnerships between industry and government. The state of the technology base is weak and desperately needs rejuvenation. It is not enough to throw money at the problem; the entire enterprise needs to be re-focused on rebuilding and revitalizing the technology base.

Fifty years from now, spacecraft will look as alien to us now as the James Webb Space Telescope might look to those who built Explorer-1 and Vanguard in the 1958. Investing in research and technology, and developing flight programs that push the frontier, will ensure that the U.S. will be the one building that spacecraft.

Achieving Cheap Access to Space, the Foundation of Commercialization

Charles M. Miller

Cheap access to space, or CATS, is a critical national and economic security objective for the coming decades. For America to dominate space in the 21st century, there is one – and only one -- capability that matters: cheap access to space. America needs a proven and pragmatic strategy to achieve cheap access to space. American free enterprise is working in space. Let's build off America's strength: we are the land of free enterprise innovation. Our national strategy should amplify and reinforce natural market forces.

We are in the middle of a paradigm shift in the space industry. A paradigm of public-private partnerships is taking the place of the old Apollo paradigm. This "new" approach is quintessentially American and actually quite old, trusted, and proven. It also mirrors the model implemented by the National Advisory Committee on Aeronautics (NACA), which helped America recapture world leadership in aviation in the 1920s and 1930s.

Alfred T. Mahan, a visionary sea power theorist from the late 19th century, identified the link between power and free enterprise, concluding, "If sea power be really based upon a peaceful and extensive commerce, aptitude for commercial pursuits must be a distinguishing feature of the nations that have at one time or another been great upon the sea."[41]

---

[41] Alfred Thayer Mahan, *The Influence of Sea Power Upon History,* 1660-1783, (Boston: Little Brown and Company, 1890: 50.

In the 21st century, an aptitude for commercial space is a distinguishing feature of nations, that are, or will be, great in space. The corollary is also true — the nation that dominates future commercial space markets will accrue great advantage to its national security, as well as great wealth for its people.

America's Future in Space Depends on CATS

Whichever country achieves cheap access to space first will start a virtuous cycle that will deliver tremendous economic and security benefits to that nation. This nation will dominate the carrying trade and create new markets, which will drive new technologies and new capabilities, which will increase flight rates and lower launch costs, which will allow that nation to expand into more new markets. Thus a virtuous cycle begins. CATS is the connection between space commerce and space power.

CATS is a fundamental requirement for expanding human civilization across the solar system. CATS implies at least an order of magnitude reduction in costs (and increase in reliability) over what is currently available. With a radical reduction in launch costs, and the resulting much higher flight rates, the United States will be a next generation space industrial power. Without it, we are not.

CATS is the key to incorporating the solar system into our economic sphere. With CATS, entirely new industries become possible. Cheap access to space is critical to closing the business case for a commercial space station. With CATS, we will see global point-to-point package delivery markets and an expansion of space adventure travel (or space tourism) to the Moon and beyond. With CATS, we will see entirely new ways of developing spacecraft, as well as on-orbit satellite assembly, repair, and servicing industries. With CATS, which will provide frequent, predictable access to space, we will see a significant increase in microgravity research and perhaps entirely new industries that nobody has predicted. With CATS, we could see space solar power become a reality in small niche power markets that are part of the multi-trillion-dollar energy industry, but are large markets by space industry standards. With CATS we will see entirely new concepts that flourish, including on-orbit propellant delivery, propellant depots, reusable space tugs, orbital transfer vehicles, tethers, and *in situ* resource utilization. With CATS, many small experiments with new technologies and concepts will suddenly be economically feasible, and we will see a renaissance in innovation, creativity, and invention in space. Combined with the other new capabilities that will emerge because of the existence of CATS, non-terrestrial resources will become economically feasible to use -- and highly valuable.

CATS can also revolutionize space science. With less expensive and more frequent launches, large spacecraft crammed with instruments can be replaced with networks of smaller spacecraft. Such an approach will enable missions that are not only less expensive, but also more redundant: a spacecraft failure will mean the loss of a single instrument, not the entire mission. This approach could radically reshape studies of the Earth, Moon, Mars, and other destinations in the solar system. Even missions that require larger monolithic spacecraft will benefit from more reliable and less expensive launch services. CATS will enable the cost-effective development of scientific spacecraft far larger than we have today — creating instruments with apertures far larger than the James Webb Space Telescope — after a commercial on-orbit assembly industry grows up around CATS.

National Security Benefits of CATS

In January 2001, the Commission to Assess United States National Security Space Management and Organization concluded that we are in danger of a "Space Pearl Harbor." Noting America's extensive dependence upon space assets, and their vulnerability to attack, the Commission concluded:

> United States deterrence and defense capabilities depend critically on assured and timely access to space. The U.S. should continue to pursue revolutionary reusable launch vehicle technologies and systems even as the U.S. moves to the next generation of expendable launch vehicles.
> Providing active and passive protection to assets that could be at risk during peacetime, crisis or conflict is increasingly urgent. ... reusable launch vehicles are needed to improve the survivability of satellites on orbit as well as the ability to rapidly replace systems that have malfunctioned, been disabled or been destroyed.[42]

The risk of a Pearl Harbor in space has only increased since this 2001 report was published. U.S. munitions rely on space assets, most notably the Global Positioning System, for their precision. If an adversary degrades those systems, we will be "dumb," "deaf," and "blind."

Potential adversaries recognize the vast conventional military superiority of the United States. In January 2007, China demonstrated the capability to destroy a satellite in orbit. Its direct ascent anti-satellite weapons used a mobile launcher, making it is very difficult to target preemptively. North Korea, one of the most unpredictable governments on the planet, has also demonstrated both a nuclear device and ballistic missiles. Iran is following in Pyongyang's footsteps, developing both ballistic missiles and nuclear technology. It has demonstrated launch vehicle technology and is on a path to demonstrate a nuclear device.

---

[42] Commission to Assess United States National Security Space Management and Organization, *Report of the Commission to Assess United States National Security Space Management and Organization* (Washington, DC: U.S. Department of Defense, 2001): 29, accessed at http://www.dod.gov/pubs/space20010111.html.

If China, North Korea, Iran, or another enemy were to destroy U.S. assets in orbit, CATS could provide the ability to rapidly replenish the satellites and augment the capabilities that satellites provide. Commercial reusable launch vehicles (RLVs), also known as spaceplanes, are the leading candidate for achieving CATS. By their very nature, commercial RLVs will have a natural surge capability, which can be used to rapidly replenish our satellites in space.

By their very existence, true RLVs or spaceplanes will eliminate most of the benefit of a surprise attack on satellites, and, therefore, would probably never need to be used in a major war. In other words, spaceplanes are a stabilizing deterrent to a space war because they eliminate much of the incentive to attack our satellites in orbit in the first place.

Achieving CATS

A fundamental challenge to focusing again on the goal of CATS is that the U.S. has tried several times to achieve this goal and failed. Achieving CATS is difficult. Any new initiative to achieve it must address the repeated national failures (Shuttle, NASP, X-33, X-34, SLI). Instead of repeating the same mistakes, and expecting different results, our nation must address the core reasons for these failures. It is time to try a different approach. It is time to go back to the future. It is time for an NACA for space.

In a traditional commercial business, the private sector makes the investments required to create a new capability. Cheap access to space is an exception.

Almost all concepts for achieving cheap access to space require a huge investment, involve large technical uncertainties, and address future markets that do not exist yet. In combination, the technical, business, regulatory, financial and political risks are far too high for any private investor to consider, based on conventional investment rules.

These problems are similar to those the United States faced in building transcontinental transportation systems, notably, the Transcontinental Railroad and the civil/commercial aviation system. Both are shining examples of successful partnerships between the U.S. government and American entrepreneurs. Neither system could be justified as a purely commercial investment, nor did the U.S. government design, develop, and operate the railroads or airplanes.

This is often the point at which the old, stale debates arise. On one side, there are those who understand government programs and want to introduce yet another program to achieve some important national priority. On the other side are those who naturally (and correctly) resist this impulse, but then find themselves arguing against all government programs, no matter how well designed, seemingly arguing for a laissez-faire approach in all cases. The NACA public-private partnership model is neither. It is a blend of the best of industry and government and represents a third option, which has proved itself repeatedly in American history.

If left alone, CATS will eventually emerge from pure market forces. The same would have been true for the Transcontinental Railroad and for 5-hour airplane flights between San Francisco and New York. Eventually private industry would have built a railroad to tie North America together. Eventually private industry would have built airplanes with the ability to cross the Atlantic, or the vast American continent, in one flight.

However, our nation cannot afford to wait that long. CATS is critical to the security of our nation. CATS is critical to the security of the free world.

The Problem with the "Big Program" Mindset

History demonstrates that the "Big Government Program" mindset almost always leads to a strategy of "Pick the Solution," which often leads to failure where the criteria for success include low-cost, reliability, and dependability. This does not mean that the next attempt for central-government-led reusable launch vehicle (RLV) program will fail, but we are now 0-for-4.

In the 1970s, NASA set out to design and build a "National Space Transportation System," also known as the Space Shuttle. NASA told the White House Office of Management and Budget in 1971 that the Shuttle would fly 50 times per year at a cost of $10.5 million a flight and deliver 65,000 pounds per launch. This equates to an estimate of $162/pound (FY1971) or $731/pound (in 2012 dollars). America spent over $30 billion (2012 dollars) to develop the Shuttle, and it failed to produce anything close to the results that were promised. In fact, the Shuttle proved to be over an order of magnitude higher in cost, as well as an order of magnitude less reliable, than promised.

In 1984, America initiated a second major effort to achieve CATS, which was called the X-30 National Aerospace Plane Program (NASP). NASP was extremely technically challenging, requiring breakthroughs in at least six distinct technologies. In a December 3, 1992 report to Congress, the Government Accounting Office (GAO) reported that the NASP program cost had grown by more than a factor of four from the 1986 estimate of $3.3 billion in 1992 dollars ($5.0 billion, 2012 dollars) to as much as $14.6 billion ($22 billion, 2012 dollars). After 10 years of activity, the NASP program was cancelled.

The U.S. did not wait long after NASP to start the next major program focused on CATS. In January 1995, NASA announced the X-33 program. NASA chose the most technically risky of the three leading concepts rather than the concept that was most likely to succeed at achieving a true RLV capability. In one of those not-so-funny situations, the civil servants responsible for serving our nation decided that more difficult technology was better, and they gave a higher score to the concept that was technically more challenging. This was a huge mistake, as neither NASA nor Lockheed Martin were willing to pay for cost overruns created by the ensuing X-33 technical problems. It was also a mistake to make a single-stage-to-orbit RLV the goal, when we had not yet developed the easier two-stage-to-orbit RLV. In an August 1999 report to Congress, the GAO estimated that the total cost to the US taxpayers for the X-33 was $1.23 billion ($1.63 billion in 2012 dollars). The X-33 RLV program was shut down in 2001.

More than a decade later, we have not yet (officially) tried again. This is not because advocates of RLVs have stopped trying. From 2002–2004, starting right after the X-33 was cancelled; NASA and the DOD combined forces to form the National Aerospace Initiative (NAI). The NAI proposed to meet all the launch requirements of DOD and NASA, thereby repeating part of the history of the Shuttle, and the total cost estimates approached $50 billion. The NAI collapsed soon afterwards. The fundamental reason for the repeated failure of these programs has to do with the limits of knowledge as well as the blind spots everyone has.

The Smart Person's Disease: The Unknown Unknowns

It is not what you know that will get you. It is not even what-you-know-that-you-don't-know. It is what-you-don't-know-that-you-don't-know. These mental blind spots are created by the fundamental nature of information and knowledge.

We create mental models based on what we know, which is based on our previous experience. The curse is that the smarter we are, and the more we know, the more likely that we will become confident, and then arrogant, about what we know.

In *Thinking, Fast and Slow*, Daniel Kahneman discusses the research of Phillip Tetlock.[43] Tetlock interviewed 284 experts who made their living on political and economic trends, and then measured the accuracy of their predictions. Kahneman summarizes:

> The results were devastating. The experts performed worse than they would have if they had simply assigned equal probabilities to each of the three potential outcomes. In other words, people who spend their time, and earn their living, studying a particular topic produce poorer predictions than dart-throwing monkeys who would have distributed their choices evenly over the options.[44]

While Tetlock's research showed that those who knew more had a slightly better ability to forecast than those who knew less, it also demonstrated that those with the most knowledge were often the least effective. Kahneman concludes, "The reason is that the person who acquires more knowledge develops an enhanced illusion of her skill and becomes unrealistically overconfident."[45] In other words, when you become confident about your knowledge, your blind spot increases in size. When you become arrogant, you stop listening, and your blind spot becomes larger. In reality, there is not much difference, if any, between confidence and arrogance.

In *The Black Swan: The Impact of the Highly Improbable*, Nassim Nicholas Taleb writes:

> Let us examine what I call epistemic arrogance, literally, our hubris concerning the limits of knowledge. Episteme is a Greek

---

43 Daniel Kahneman, *Thinking, Fast and Slow* (New York: Farrar, Straus and Giroux, 2011).
44 Kahneman, *Thinking, Fast and Slow*, 219.
45 Kahneman, *Thinking, Fast and Slow*, 219.

word that refers to knowledge; giving a Greek name to an abstract concept makes it sound important. True, our knowledge does grow, but it is threatened by greater increases in confidence, which make our increase in knowledge at the same time an increase in confusion, ignorance, and conceit.[46]

Everybody agrees that achieving CATS is critical for the nation. Everybody agrees that we should marshal the efforts of the best and brightest people in the nation to tackle this challenge. One problem is that we don't know who the best and brightest people really are. A second problem is that we really can't know what will be the best solution when we develop a plan to attack the problem. A third problem is that we get too specific with lessons learned.

The Problem with Specific Lessons Learned

Many have studied the failures of the Shuttle, NASP, and X-33 for lessons to apply in the next attempt build RLVs, and rightly so. The managers of NASP learned from the failures of the Shuttle, and the managers of the X-33 program learned from the failures of both the Shuttle and NASP. While there is much to learn from each failure to achieve CATS, the space industry is dominated by engineers who have a tendency to focus on specific engineering challenges that are easier to quantify. Other critical issues that are harder to measure, and more difficult to understand, are either missed or forgotten.

Taleb continues:

> Another related human impediment comes from excessive focus on what we do know: we tend to learn the precise, not the general.

---

[46] Nassim Nicholas Taleb, *The Black Swan: The Impact of the Highly Improbable* (New York: Random House Trade Paperbacks, 2007): 138.

What did people learn from the 9/11 episode? Did they learn that some events, owing to their dynamics, stand largely outside the realm of the predictable? No. Did they learn the built-in defect of conventional wisdom? No. What did they figure out? They learned precise rules for avoiding Islamic prototerrorists and tall buildings... The story of the Maginot Line shows how we are conditioned to be specific. The French, after the Great War, built a wall along the previous German invasion route to prevent reinvasion—Hitler just (almost) effortlessly went around it. The French had been excellent students of history; they just learned with too much precision.

We do not spontaneously learn that we don't learn that we don't learn.[47]

The Problem with the Laissez-faire Mindset

Those who promote a purely laissez-faire approach have just as much epistemic arrogance as those who promote yet another big government program. Both sides tend to ignore the hard empirical data that is inconsistent with their views.

The empirical data is clear: the laissez-faire "just leave us alone" approach has not succeeded at achieving CATS either. A series of companies, with names like Pacific American, Kistler, Pioneer Rocketplane, Kelly Space & Technology, Rotary Rocket, and many others have tried to privately finance totally reusable launch vehicles (RLVs) over the last three decades. Many of these companies have had brilliant concepts, led by brilliant engineers, but the size of the investment required to fully develop their systems has proven beyond their reach. While several of these firms later asked for, and received, funding from the government, they all started with a purely commercial business plan, and they all failed.

---

47 Taleb, *The Black Swan: The Impact of the Highly Improbable*, xxv.

Traditional investment markets have sent a clear signal — the business plans for conventional RLVs do not close. Conventional investors from Wall Street, to Silicon Valley, to corporate board rooms, have repeatedly taken a pass on RLV investments based on their assessment of risk-adjusted return on investment and the large package of business, market, technical, and financial risks.

100 Years of Empirical Data, not just 40

The U.S. Government's first attempts to invent flight over a century ago demonstrate a striking similarity with its repeated failures to achieve cheap access to space. In the immortal words of Yogi Berra, it will be "Deja vu all over again."

Lord Kelvin, President of the Royal Society, summed up the conventional wisdom on flight at the end of the 19th century when he declared: "Heavier than air flying machines are impossible." In spite of this, the U.S. Department of War decided to fund a $50,000 program by Dr. Samuel P. Langley to develop a practical human flying machine. An advisory committee, with two officers from the Navy, two from the Army, and a mathematician from the US Naval Observatory, concluded that Langley's project could help with wartime reconnaissance, "communication between stations isolated from each other by ordinary means," and serving as "an engine of offense."[48]

It was logical for the U.S. Government to hire the best and brightest program manager it could find. By all reasonable criteria, Dr. Langley was that person. James Tobin, a historian of the period, writes: "In public stature and prestige, Langley was the most prominent scientist in the United

---

48 Langley Scrapbook, 1897-98, RU 7003, Smithsonian Institution Archives.

States. His best friend was Alexander Graham Bell, the inventor of the telephone. Langley was a frequent guest at the White House."[49]

Dr. Langley was a brilliant visionary who dedicated the later years of his life to inventing the airplane. After becoming Secretary of the Smithsonian in 1886, he immediately instituted a basic research program on flight using the Smithsonian's resources. In 1896, following ten years of sustained experimentation, Langley finally proved that human flight was achievable when his 16-foot unmanned flying machine, powered by a steam engine, flew nearly one mile.

Langley then set out to raise $50,000 (in then year dollars) to take his experimental research to the next level. It took two years for his fundraising campaign to succeed. He would soon direct a full-time staff, which reached about 10 people. Langley told the War Department with "confidence" that "the machine will be completely built and ready for trial within a year" — meaning by 1899.[50] Langley would spend the entire $50,000 provided by the War Department, run out of money, and then raise another estimated $20,000 from other sources — and take five years until the first flight test in 1903.

Dr. Langley made two flight attempts. The tests were open to the public and the media, and two members of Congress were explicitly invited. The first attempt failed on October 7, 1903. Since the vehicle incurred only minor damage, a second attempt was possible. It, too, failed on December 8, destroying the craft. After the second public failure, the project became known as "Langley's Folly." Both the media and members of Congress harshly attacked Langley and the Department of War's waste of public funds. The *New York Times* editorialized:

---

49 James Tobin, *To Conquer the Air: The Wright Brothers and the Great Race for Flight* (New York: The Free Press, 2003): 4.
50 Langley Scrapbook, 1897-98, RU 7003, Smithsonian Institution Archives, "Extract from Proceeding of Board of Ordnance and Fortification".

> The ridiculous fiasco which attended the attempt at aerial navigation in the Langley flying machine was not unexpected… it might be assumed that the flying machine which will really fly might be evolved by the combined and continuous efforts of mathematicians and mechanicians in from one million to ten million years.[51]

Both the Smithsonian and the Department of War lost their appetite for continuing. The Smithsonian's regents forbade Langley from continuing his research. The Department of War concluded, "We are still far from the ultimate goal [of human flight]."[52] This was an unfortunate ending for a great American.

Hindsight is 20/20, so it is easy for us to see that Dr. Langley was not America's best and brightest on the subject of inventing flight. But nobody knew this at the time. If you put yourself in the position of the Department of War in 1898 — if you had already decided to fund a centrally-controlled R&D program to design and build a practical human flying machine — the most rational and logical course of action was to hire Dr. Langley to be the program manager.

How could anybody know that America's best and brightest on the subject of flight were two brothers who owned a bicycle shop in Dayton, Ohio? Not even the Wright Brothers knew. This is clear from the following letter from Wilbur Wright, who wrote to the Smithsonian on May 30, 1899, asking for information:

> Dear Sirs,
>
> I have been interested in the problem of mechanical and human flight ever since ever since as a boy I constructed a number of bats [toy planes – Editor] of various sizes… My observations

---

51 See copy of full NY Times article at http://www.freerepublic.com/focus/f-news/1042349/posts
52 Charles Gibbs-Smith, *Aviation: An Historical Survey From Its Origins to the End of World War II* (London, HMSO, 1960): 67.

since have only convinced me more firmly that human flight is possible and practicable. It is only a question of knowledge and skill as in all acrobatic feats… I am an enthusiast, but not a crank in the sense that I have some pet theories as to the proper construction of a flying machine… I am about to begin a systematic study of the subject in preparation for practical work to which I expect to devote what time I can spare from my regular business. I wish to obtain such papers as the Smithsonian Institution has published on this subject, and if possible a list of other works in print in the English language… I wish to avail myself of all that is already known and then if possible add my mite to help on the future worker who will attain final success.

Yours truly

Wilbur Wright[53]

If even Wilbur Wright did not know what was to come, how could anybody know that a couple of American entrepreneurs would do what most thought was impossible — invent the first practical airplane — and do so in their spare time as a hobby? How could anybody know that these two brothers, who did not graduate from high school, would take just four years to overcome a combination of vexing problems — highly-efficient propellers, effective lift, and (most importantly) active dynamic control — which the esteemed Dr. Langley would fail to solve over a seventeen-year period?

It was not knowable. In sum, Dr. Langley spent an estimated $70,000 trying to invent a practical flying airplane, and failed. The Wright Brothers, working part-time over four years, spent an estimated $1,000, and succeeded. The War Department's mistake was not in its specific choice of Dr. Langley to manage the program: it was in choosing to organize

---

[53] Marvin Wilks McFarland, editor, *The Papers of Wilbur and Orville Wright, Vol. 1* (New York: McGraw-Hill Book Company, 1953): 4-5.

a big government program that could only afford to make one specific attempt to solve the problem. The War Department's paradigm in 1898 was to choose "The Solution." When they failed, the entire initiative came to a stop. Dr. Langley's noble attempt was the equivalent of the repeated (noble) failures by the Shuttle, NASP, X-33, and the National Aerospace Initiative to achieve cheap access to space.

The Wright Brothers — The Quintessential American Entrepreneurs

The city of Dayton, Ohio was founded in 1805, and in some respects, the Dayton culture of 1900 was reminiscent of Silicon Valley and Seattle. The frontier had only recently passed through Ohio, and it was a city where new ideas flourished. By 1900, Dayton had become a center of manufacturing and invention, a "city of a thousand factories," and had the highest number of patent filings on a per capita basis in the nation.[54]

From 1899 to 1903, the Wright Brothers would devote themselves to the hobby of a systemic study of human flight. They took a fundamentally empirical approach: they wanted to maximize their flight time, as the amount of learning from actually flying something was much greater than theorizing about flying. This was an early version of the rapid prototyping that is used today by companies like Scaled Composites and many firms in Silicon Valley, and was used in the DC-X program. One key method of the Wright Brothers was to quickly test their ideas, which some today call "build a little, test a little." When Wilbur had his first idea for roll control, he decided to quickly test it by building a small kite. The kite looked much like a box, with two planes that were 5 feet by 13 inches, and about 13 inches apart. With two sticks tied to strings he could warp the kite wings. In July 1899, Wilbur Wright took the kite out on a windy day in small field in West Dayton. It worked.

---

54 Mark Bernstein, *Grand Eccentrics: Turning the Century: Dayton and the Inventing of America* (Wilmington, OH: Orange Frazer Press Inc, 1996): 15-18.

Less than two months after contacting the Smithsonian, Wilbur Wright surpassed Langley – as well as everybody else in the world -- in his fundamental understanding of aeronautics. Wright not only came up with the fundamentally new idea of roll control, but he invented a mechanism for roll control and tested it in a small kite in downtown Dayton. John D. Anderson Jr., the curator for aerodynamics at the Smithsonian Institution and the Glenn L. Martin Distinguished Professor Emeritus in the Department of Aerospace Engineering at the University of Maryland, writes: "The Wrights' most important and original contribution to the technology of flight occurred very early in their aeronautical development program—the appreciation of the need for lateral control and a mechanical mechanism for achieving it."[55]

For a period of four years, the Wrights made rapid and continual advances, based on executing rapid-do-learn loops as well as their commitment to basic research and hard data. It was their innovative methods combined with their genius that produced the breakthrough in flight. Their entrepreneurial approach is fundamentally similar to the American entrepreneurial culture today, exemplified in Silicon Valley and many of the New Space firms. Unfortunately for American leadership in the early part of the 20th Century, this innovative entrepreneurial culture was also alive and well in a different country — France.

America Learns Wrong Lesson & Becomes a Follower

What was the lesson learned by America in 1908-09 after the Wright brothers shocked the world with their demonstration of practical, sustained flight? American entrepreneurs had beaten the U.S. government and the central program mindset. After buying the first Wright airplane, which it was contractually obligated to do, how much funding did the U.S.

---

[55] John D. Anderson, Jr., *The Airplane: A History of Its Technology* (Reston, VA: American Institute of Aeronautics and Astronautics, 2002): 88.

government allocate to buying additional airplanes to take advantage of this clear revolution in technology?

None. After providing convincing proof to the Army in 1908 that the era of heavier-than-air aircraft was upon us, the U.S. Government did not purchase a second plane until 1911. This slow pace was not for lack of trying by the U.S. Army Signal Corps. General Allen, Chief Signal Officer of the U.S. Army, requested $200,000[56] for aeronautics for fiscal year 1908, based on Orville Wright's 1908 flight demonstration. According to *Aviation Online Magazine*, General Allen "repeated this request for 1909 and again for 1910. Congress failed each time to approve any funding."[57] The U.S. Army actually leased its second airplane, a Wright "Type B" from Robert Collier, in February 1911. The lease price was $1.00 per month.[58] The nascent U.S. Army air force had been reduced to the equivalent of begging. The first congressional appropriation for Army aviation was $125,000 for fiscal year 1912.[59]

In 1908, while Orville Wright was demonstrating their technology for the U.S. Army, Wilbur demonstrated it in Paris. France, Britain, Russia, and Germany immediately recognized the implications and accelerated their nascent basic research and programs to buy planes from industry. Within one year, French aviation entrepreneurs had caught up to the Wright brothers, and within a few years all of Europe had blown past America's best. French entrepreneurs invented many critical aviation technologies, including the fuselage, the wheeled landing gear, the aileron, the rotary engine, and the stick and rudder control system. When added to the Wright brothers' innovations, the French would quickly capture world leadership.

---

56 War Department, *Annual Report, 1908, Volume 2* (Washington, DC: U.S. War Department, 1908): 214.
57 Aviation Online, "Signal Corps No. 1: Purchasing and Supporting the Army's First Airplane." *Aviation Online*, accessed at http://avstop.com/history/signal.htm.
58 William Sherman, *Air Warfare* (Honolulu, HI: University Press of the Pacific, 2004).
59 National Park Service, www.nps.gov/nr/travel/aviation/col.htm.

Historian John Morrow writes: "A critical factor in the rise of military aviation from 1909 through 1911 was the rapid improvement of aircraft performance effected by the French, as they raised the speed record from 48 mph in 1909 to 83 mph in 1911. Distance also increased from 146 miles to 449 miles, and height increased from 1,468 feet to 12,828 feet."[60]

By the beginning of World War I in 1914, France would become the indisputable world leader in aviation. If aerospace leadership is a worthy objective, and if public policy is worthy of serious study, then we must ask "What exactly was going on with French aviation in the first and second decades of the 20th Century?"

The French Aviation Miracle

France had many aviation enthusiasts who were contemporaries to the Wright brothers. Where America had two serious aviation entrepreneurial teams, composed of the Wright brothers and Glenn Curtiss, France had many more and was an intense hot-bed of entrepreneurial activity. The majority of aviation activity in France from 1900-1909 was privately financed. In parallel, there was an active public movement to promote investments in aviation in France, most notably by the Aero-Club De France and Ferdinand Ferber.

French aviation was an intense competition between teams and companies led by entrepreneurs such as Alberto Santos-Dumont, Gabrielle and Charles Voison, Robert Esnault-Pelteri, Louis Breguet, Clement-Bayard, Jules Gastambide, and Louis Blériot. Alberto Santos-Dumont was the son of a wealthy coffee-plantation owner, and would spend part of his inheritance developing and flying airplanes. Robert Esnault-Pelteri would invent the joystick and independently invent the aileron. He almost bankrupted his industrialist father, but recovered somewhat based on his

---

60 John Morrow, *The Great War in the Air: Military Aviation from 1909 to 1921* (Washington, DC: Smithsonian Institution Press, 1993).

joystick patents. Adolphe Clément-Bayard made his initial money building automobiles, which he then invested in aircraft, including the world's first aircraft assembly line in 1908. Jules Gastambide was another industrialist who made his initial fortune building powerful lightweight engines for automobiles and boats, which were then sold as the advanced Antoinnette V8 aircraft engine. The Antoinnette company would then start building innovative and highly competitive aircraft. Ernest Archdeacon and Henri Deutsch de la Meurthe would fund a series of aviation prizes using their private wealth. The Voison brothers inherited their grandfather's wealth, and raised additional private investment. Gustav Eiffel invested part of his private fortune by turning his famous tower into an aeronautical research platform in 1905, where he proved that air resistance was proportional to velocity squared.

The story of Louis Blériot illuminates what was taking place in France during this time.[61] In 1895, at the age of 23, Blériot founded a company to build headlamps and accessories for automobiles, which was an emerging transportation revolution. He was fresh out of college and the classic entrepreneur. By the beginning of the 20th century he was earning an average of 60,000 francs a year, which he used "to endulge (sic) in his first aeronautical experiments" starting in 1900. In 1905 he watched Gabriel Voison test fly an experimental float glider that was produced for Archdeacon. Blériot ordered a new and revised version on the spot, which Voison built and tested in a month. This second float glider was unstable and crashed into the water — but instead of despairing from the failure — Blériot decided to get serious. He then financed the Blériot-Voison company, which became the first company in the world founded solely to produce airplanes. A year later, in 1906, they would test fly the

---

61 The following is drawn from Thomas Crouch, *Bleriot XI: The Story of a Classic Aircraft* (Washington, DC: Smithsonian Institution Press, 1982).

Blériot IV in attempt to win the Aero-Club de France prize of 1,500 francs for the first flight of over 100 meters. They would lose to Santos-Dumont, who won the prize in November 1906. The Blériot-Voison partnership dissolved, but they each continued independently. Blériot intensified his efforts, again, by setting up multiple independent teams that would construct prototypes in parallel. Starting in 1907, he was testing two new prototypes a year based using an empirical process with rapid do-learn-loops. In 1907, both Robert Esnault-Pelteri and Blériot would file for a patent on versions of the modern stick and rudder control within a month of each other.

When the Wright brothers showed up in Paris in 1908 to fly their system, they clearly demonstrated their world leadership, primarily due to their fundamental insights on dynamic stability. All of France acknowledged this — from French aviator Leon Delagrange who remarked "Nous sommes battus" (We are beaten) to Bleriot who was quoted in the *New York Herald* saying, "It is Marvelous."[62] However, the three years, from 1905-1908, which the Wright Brothers spent trying to sell their Wright Flyer and making zero progress in aeronautical research, had a huge opportunity cost. The best use of their time would have been in designing the next generation of airplane. During this period the French independently developed and flight-tested the aileron, and invented the stick-and-rudder control system, the fuselage, the wheeled landing gear, and implemented the tractor-pulling monoplane.

Finally, where the Wrights were not used to competition, the French aviation entrepreneurs were highly competitive. They would quickly borrow and adopt ideas from each other, which is eerily familiar for anybody who follows the culture of modern day Silicon Valley. They also borrowed

---

62 Charles Harvard Gibbs-Smith, *The Rebirth of European Aviation, 1902-1908: A Study of the Wright Brothers Influence* (London: Her Majesty's Stationery Office, 1974): 288.

from the Wright brothers. Within a year, they would catch the Wright brothers and then leap ahead.

While private financiers primarily drove early French aviation, a shift began in Europe after the 1908 Wright demonstration. This shift accelerated in 1909 after the Blériot channel crossing. Clement Ader, a French aviation entrepreneur in the late 19th century, amplified the shift by publishing *L'Aviation Militaire*, which described some of the military implications in detail. It was so popular that it was printed in 10 editions from 1909 to 1914.

The French government began to aggressively stimulate its entrepreneurial industry by purchasing large numbers of airplanes from those commercial firms. In April 1910, the French government purchased 35 aircraft from just one company, Voison.[63] Tom Crouch, from the Department of Aeronautics at the Smithsonian Institution, writes:

> In 1910-1911, a period during which the U.S. Army took delivery of 14 airplanes, the French government ordered over 200 flying machines. Across the face of an increasingly troubled Europe, success in the air symbolized the courage and strength of the nation. That was particularly true in France, where a growing number of citizens were determined that the nation which had sent the first human beings aloft should, at any price, retain leadership in the air. Proof of the extraordinary level of popular enthusiasm came in 1912, when the National Aviation Committee raised four million francs with which to supplement the national budget for military aviation.[64]

But it was more than just government funding. The environment and culture for aviation across Europe was more competitive than in the

---

63 John Morrow, *The Great War in the Air*, 33.
64 Tom Crouch, *Blaming Wilbur and Orville: The Wright Patent Suits and the Growth of American Aeronautics* (Dallas, TX: University of Texas at Dallas, 2003).

United States. There were more and larger privately funded prizes and more intense competitive pressure driven by more firms. Crouch explains:

> Europe not only offered more contests and richer prizes, it provided a much higher level of competition. In the U.S. the leading aviators were members [of] two or three touring exhibition teams who earned salaries for performing aerial stunts to thrill crowds of paying customers. There were no better pilots in the world than men like Lincoln Beachey and Walter Brookins, but they had not been tested under the constant pressure to fly higher, faster and farther against a wide range of competitors, week after week. More important, their technology had not been tested either.
>
> With little incentive for change, American builders like Glenn Curtiss and Glenn Martin remained largely committed to the original configuration of the Wright airplane — a pusher biplane with a canard elevator — until 1910-1911.
>
> Strenuous competition between a relatively large number of designers and aviators in Europe led to the exploration of a wide range of configurations, the use of new materials, and improved control systems and power plants[65]

Even after the U.S. Government started appropriating funds to buy airplanes in 1911, other countries were now leaping further ahead of the United States in their commitment to dominating this strategically critical technology. Crouch reports that other countries were spending an order of magnitude more than the United States:

> As early as 1912, the Secretary of the Navy pointed out that the U.S. lagged far behind other leading nations of the world in expenditures for aeronautics. France, he estimated, had spent $7,400,000 on flight to date. Russia was in second place, with an

---

[65] Tom Crouch, *Blaming Wilbur and Orville*.

> expenditure of $5,000,000, followed by: Germany, $2,250,000; and Great Britain and Italy, $2,100,000 each. Even Japan ($600,000) had out spent the U.S. ($140,000).
>
> ... By 1914, France was, by almost any measure, the world's leading aeronautical power. While French government policy was neither entirely consistent nor completely rational, political decisions were, as historians Emmanuel Chadeau and John Morrow have noted, primarily responsible for creating the strongest aviation industry in Europe.[66]

It was not only the amount of money that France committed but also how they spent the money. France's strategy was to amplify the entrepreneurial forces in its country, to maximize competition, and to stimulate new entrants.

In a history of early aviation sponsored by NASA, Alex Roland writes:

> A small group of like-minded men in the United States found it a national embarrassment—not to say a danger—which the country where aviation began should trail so far behind the Europeans. They saw aviation as an infant Hercules with boundless potential for national defense, commerce, and even melioration of the human predicament... They wanted to see the United States lead in every phase of aviation, and they believed... that the advance of aeronautics would come with scientific research. They also wanted to see larger budgets for military aviation, the encouragement of commercial aviation, and the nurturing of an aircraft-manufacturing industry, they wanted first and foremost a national aeronautical research laboratory to rival those in Europe.[67]

---

66 Tom Crouch, *Blaming Wilbur and Orville*.
67 Alex Roland, *Model Research: The National Advisory Committee for Aeronautics, 1915-1958* (Washington, DC: NASA, 1985).

The Taft Administration was persuaded by the organized campaign of these advocates to recommend the creation of a federal aeronautical research laboratory. Taft's proposal was modeled after the British Advisory Council on Aeronautics, but Congress voted it down in 1913. Fighting among the various interests who wanted to be in charge of the effort was the root cause of the delay. It took an irrefutable national security crisis to cause Congress to act. World War I broke out in 1914, and in March of 1915 Congress enacted Taft's bill.

Even so, by the time the United States entered the war, it was so far behind Europe in flight that it would have been suicide to send our young aviators into the air against European planes. America had become a distant follower in aviation technology, and this had national security consequences. The United States was forced to license and build French-and-British-designed planes. The Dayton-Wright Company would end up manufacturing 3,604 British De Havilland DH-4s for the war.

Lessons Learned from Early French Aviation

The conclusion is clear. France's public-private partnership, designed to stimulate and accelerate market-based entrepreneurial innovation, trumped the more laissez-faire American approach. The French aviation miracle was founded upon the innovation of the French aviation entrepreneurs, amplified by a national strategy to stimulate and reinforce market forces, and honed in a highly competitive environment.

Public private partnerships (PPPs) represent a third way. They are not purely commercial, nor purely government, and do not fit easy definitions. Today, Washington ties itself in knots arguing over "what does 'commercial space' mean?" Multiple NASA administrators, as well as space policy professionals, struggle with and argue over this question. While versions of what "commercial" means have been defined in law, and more recently in national space policy, arguments still break out among policy professionals about what it means.

The desire for clear definitions, leading to complete agreement by all parties, is almost certainly futile. In the meantime, we need to focus on what works in order to achieve the critical national goal of cheap access to space. We have a good working definition in law and policy, and good historical models for PPPs. The key to an effective partnership is effectively assigning clear roles and responsibilities to each party, according to their skills, and aligning their interests so they are all pulling in the same direction. Fortunately, there is no real debate over whether the NACA partnership model works. It worked for many decades, and it will work again. History provides the proof.

Lessons learned from the early years of the NACA

From the outset, the NACA's core mission was to solve "practical problems" related to flight. Conventional wisdom is that the NACA produced groundbreaking research with its wind tunnels, producing significant drag reduction in all vehicles, the low drag engine cowling, de-icing technology, new airfoil designs, and the variable pitch propeller.

But history forgets the early years. The NACA would not have a wind tunnel until five years after its creation. It had very little money during its first decade, but what it accomplished with little money is amazing. The key was that it saw its unofficial mission as stimulating and energizing a new industry. According to Roland,

> ...The members of the NACA believed to a man that the future of aviation in the United States depended on a healthy and prosperous aircraft-manufacturing industry, and that it was the NACA's duty to help where it could. From the outset, the NACA

was an industry booster limited only by its need to be fair and impartial in disbursing favors and assistance.[68]

While the NACA was created in 1915, and construction of the wind tunnels at Langley would begin in 1917, it would take until 1920 for the first wind tunnel to become operational. This forced the NACA leadership to figure out how it could help the industry with very little, and sometimes zero, funding.

The NACA took a holistic approach to addressing the challenges of stimulating a healthy and prosperous American aircraft industry. The NACA was structured to bring all key national players to the table, including the Army, Navy, Smithsonian, Bureau of Standards, and Weather Bureau. Therefore, one of the NACA's strengths was its ability to solve coordination and cooperation problems.

It was clear that industry needed customers for airplanes. Starting in 1916, the NACA advocated the creation of airmail services, and the Army began these services in 1918. The Kelly Airmail Act of 1925 was passed to create a demand-based subsidy for commercial airmail carriers. Significant credit must go to the NACA as the U.S. government's leading aviation policy advisor. The Kelly Airmail Act was the historical predecessor to NASA's International Space Station Commercial Resupply Services contracts. In both cases, the U.S. government provided a demand-side stimulus for a clearly established need that helped U.S. industry close a business case, justifying the investment of private capital.

In 1916, the NACA played a critical role in creating the specifications of the Liberty engine. As World War I broke out, America was a world leader in internal combustion engine technology and production. All parties wanted a world-class American aircraft engine, but they could not agree on what that engine should look like. The Department of War

---

68 Alex Roland, *Model Research*.

blamed the engine manufacturers for giving it conflicting approaches. The engine manufacturers blamed the Department of War for not making up its mind and establishing requirements. This stalemate was exacerbated by a law creating barriers to government cooperative activities with industry.[69] The NACA had unique powers in this area, allowing it to bring together all key parties and establish the top-level specifications of the Liberty Engine. A year after the NACA interceded, the first Liberty engines starting rolling off U.S. assembly lines in 1917. British authorities were so impressed that they asked America to stop sending aircraft and to just send over Liberty engines.

In early 1917, the Secretaries of the Navy and War asked the NACA to intercede again to end the destructive patent fight between the companies that owned the patents of the Wright brothers and Glen Curtiss. The NACA managed the process to do so — basically making the companies an offer they could not refuse: either sell your patents to a new organization that the U.S. government will create, or the U.S. government will, by law, take the patents away from you. This was an extraordinary step, but the country was at war and continued fighting by industry was therefore a national security problem. As a result, the Aircraft Manufacturer's Association (AMA) was used to pool their patents. Any American company could get a license to all the patents by becoming a member of the AMA and paying a license fee for every aircraft built in America. Thus, the NACA deserves credit for quickly ending a destructive fight that was creating barriers to new innovation.

The NACA took many other actions that had nothing to do with solving practical technical problems in order to stimulate an American aviation industry. It persuaded commercial insurers to begin insuring

---

[69] The 1910 Civil Sundry Act required congressional approval before U.S. government officials could serve on independent commissions or expend federal funds on the same. See Alex Roland, *Model Research*.

aviation. It recommended an increase in the budget for the Weather Bureau to promote safety in aviation. It developed methods for mapping from planes. The NACA also advocated the creation of a new federal agency to regulate safety in aeronautics throughout the 1920s, culminating in an official recommendation in 1925. In 1926, the U.S. Congress passed the Air Commerce Act, which created the Bureau of Aeronautics, the predecessor of the Federal Aviation Administration.

As an example of its core research function, in 1927 the NACA completed its research on the engine cowling in its advanced variable-density wind tunnel, producing a 60-percent reduction in drag and a 14-percent increase in speed. Two years later, in February 1929, after a Lockheed Vega outfitted with the NACA cowling set a new transcontinental record, Lockheed wired "Record impossible without new cowling. All credit due NACA for painstaking and accurate research."[70]

Commercial airplane innovation was increasingly driven by demand from commercial airliners to transport people across a continent, but it was still dependent upon government-sponsored airmail. Even after the huge growth in the demand for commercial passenger travel driven by the Lindbergh crossing of the Atlantic in 1927, airlines could not close a business case based solely on passenger travel in the early 1930s. Still, American aviation technology was rapidly evolving under this partnership — it was a highly competitive environment with multiple commercial competitors that was supported and enabled by the U.S. Government in multiple ways. In 1993, Boeing came out with its 247 while Douglas revealed the DC-2. The *London Morning Post* remarked "Preconceived ideas of the maximum speed limitation of the standard commercial aeroplane have been blown sky high ... America now has in hundreds, standard commercial

---

70 James Hansen, The Wind and Beyond: A Documentary Journey into the History of Aerodynamics in America (Washington, DC: NASA, 2009): 171.

aeroplanes with a higher top speed than the fastest aeroplane in regular service in the whole of the Royal Air Force."[71] David Cartwright reports:

> From the 1930s on, American commercial equipment was the best in the world. In 1934, more than half of American passengers flew at cruising speeds over 160 mph. The European airlines combined could muster just thirty-three planes that cruised at more than 125 mph. The best European long-distance racing planes could barely outperform the new generation of American airliners.[72]

Douglas Aircraft produced the first DC-2s in 1934 as a market-based response to Boeing's attempt to monopolize the first 247s for its wholly owned subsidiary, United Airlines. The DC-3, introduced in 1936, carried 50 percent more passengers than the DC-2, for only a 10 percent increase in cost. American Airlines purchased the first DC-3s with a government loan from the Reconstruction Finance Corporation.[73] The DC-3 was the first plane that could make money on each flight without a government airmail subsidy. Demand for aircraft grew exponentially. By cutting the link to government-funded airmail, the DC-3 opened airline passenger travel to millions. Douglas sold over 10,000 planes to commercial and military customers.

What was the end result of all this public-private partnership activity in American aviation? Starting from a position that was far behind the world in 1915, America had clearly caught up within a decade of the creation of the NACA: by the time Lindbergh crossed the Atlantic in 1927, America had demonstrably caught the rest of the world. By the early 1930s, with the creation of Douglas DC-2 and the Boeing 247, America

---

71 Carroll Glines, *Roscoe Turner: America's Master Showman* (Washington, DC: Smithsonian Institution Scholarly Press, 1999): 205.
72 David Cartwright, *Sky as Frontier: Adventure, Aviation, and Empire* (College Station, TX: Texas A&M Press, 2005): 15.
73 David Cartwright, *Sky as Frontier*, 100.

had become the clear global leader in long-range aviation, which provided critical capabilities and advantages to the United States in World War II. Since that time, America has never relinquished its title as the world leader in aviation.

The NACA model works today too. A paradigm shift is underway. A public-private partnership, similar in important ways to the NACA model, is already emerging but most people have not noticed. This paradigm shift began in the Reagan administration, and continues to this day.

The creation of the Office of Commercial Space Transportation in 1984; the legal requirement for NASA to purchase commercial launch services established by the Launch Services Purchases Act in 1990, and the requirement for NASA to purchase commercial ISS cargo delivery established by Commercial Space Act of 1998 are waypoints. More recently, the Commercial Crew & Cargo program, begun by the Bush Administration and strongly supported by the Obama administration, would look familiar to the American government officials that created and supported the NACA, the Kelly Airmail Act of 1925, and the Air Commerce Act of 1926. The COTS program, and the Commercial Crew Development program, use NACA-like approaches.

The creation of the Atlas V, Delta IV, Falcon 9, and Antares launch vehicles are wonderful examples of how effective public-private partnerships can be. Each of these launch vehicles was the result of the U.S. government using its "other transactions authority" to partially fund industry-led development of a brand new launch vehicle. In each case, U.S. industry had significant private capital at risk and was responsible for development, and the U.S. government assisted as requested by industry. The result is that we are 4-for-4 in development of new U.S. launch vehicles using commercially-controlled development processes with major incentives by the U.S. government.

While neither the Delta IV nor the Atlas V has been successful in the commercial marketplace, they were a great deal for the American government. As designed, they are highly reliable expendable launch vehicles, which are lower in cost than their predecessors, and capable of achieving even lower costs if operated commercially. However, after the collapse of the big satellite constellations in the early 2000s, the large commercial opportunity disappeared. Boeing and Lockheed reduced their focus on commercial markets. At that moment, doing everything that their U.S. national security customers asked them to do made perfect sense to these industry leaders.

As discussed earlier, the U.S. Air Force and the National Reconnaissance Office spend much more money on satellites than on launch vehicles. Thus, it is economically rational to spend more for even small increases in reliability. The result is that mission assurance costs on the Delta IV and Atlas V have significantly increased. For this reason, they are now highly reliable, and very expensive, launch vehicles that have few commercial customers.

The facts are clear — America has a good record of success for developing new space systems when it uses public-private-partnerships founded on commercial processes. This must be contrasted with the series of failures to achieve CATS using the standard government-controlled program development processes. We must also remember that the immediate predecessors to the Delta IV and Atlas V were the Advanced Launch System (ALS) and National Launch System (NLS) programs, both traditional government programs that failed. The reason the DOD was willing to try using its "other transactions authority" for developing Delta IV and Atlas V was because it had given up trying to build new launch vehicles after the failure of ALS and NLS.

There is compelling proof that the public-private-partnership model works and can save the U.S. taxpayers billions of dollars. NASA's cost experts used the NASA-Air Force Cost Model (NAFCOM) to estimate what it would cost to develop the Falcon 1 and Falcon 9, plus new engines and avionics, using the traditional NASA approach. The answer was $3.9 Billion. SpaceX's actual costs were less than half a billion dollars.

This factor of eight difference has shocked many -- but this has been seen before. In the early 1990s, SpaceHab spent $150 million to design, develop, and manufacture two pressurized modules that had to be human-spaceflight-certified to go on the Space Shuttle. NASA paid Price Waterhouse to use NASA's then-standard cost model to estimate what it would cost NASA to develop a similar system using traditional methods. The answer was $1.2 billion. In other words, SpaceHab demonstrated the same savings that SpaceX demonstrated almost two decades later. While SpaceHab ultimately failed to develop true commercial demand for its services, the key point is that it saved the U.S. government money. These examples prove that commercially controlled development processes, where industry bears the risk of cost over-runs (as well as the gains from lowering costs and faster development), offer a significantly lower cost acquisition approach.

Conclusion

Cheap access to space (CATS) is the most important near-term strategic objective the United States could, and should, pursue in space. CATS is the key to opening up the space frontier and to fulfilling the many unrealized promises of space. With CATS, we will put humans back on the Moon, settle permanently on Mars, and travel throughout the Solar system. Most importantly, CATS is critical to U.S. national security throughout the 21st Century and beyond.

The NACA helped America recapture leadership for the last 100 years. With CATS as our goal, and the NACA model as our means, American leadership in aerospace will be assured for the next 100.

Conclusion: The Vision Thing
Eric R. Sterner

As discussed in the introduction, examinations of the nation's civil space program have decried its lack of a vision since the end of the Apollo program. Yet, a so-called unifying vision only guided NASA for slightly more than a decade, from Kennedy's May 1961 challenge to land a man on the moon to December 1972, when Gene Cernan left the last footsteps on our nearest celestial neighbor. As of this writing, NASA is fifty-five years old. The agency pursued a so-called unifying vision for less than a quarter of its lifespan. No wonder the buckets of ink and forests of trees expended decrying the situation and calling for a new unifying vision.

Is that view correct? Perhaps not. NASA's founding charter, the National Aeronautics and Space Act of 1958, as amended through 2010, offers a simple rationale for NASA's existence. Congress simply "declares that the general welfare and security of the United States require that adequate provision be made for aeronautical and space activities."[74] In other words, aeronautical and space activities are good things in and of themselves. If that is the case, then it does not matter so much what those activities are, be they visiting the Moon and Mars, bringing an asteroid into Earth orbit, settling the solar system, developing new technologies, or contributing directly to life on Earth. All equally serve the general welfare, albeit in different ways. In that event, the space program does have

---

74 The National Aeronautics and Space Act, Public Law 111-314, December 18, 2010. Hereafter referred to as "The NASA Act." Available at: http://www.nasa.gov/offices/ogc/about/space_act1.html.

a vision, simply one that has less to do with achieving a desired end-state and more to do with justifying a particular level of effort. The lack of a so-called unifying vision is not a weakness, so much as a reflection of the myriad ways in which space activities contribute to the general welfare.

The NASA Act appears to embrace that mindset. It lists more than a dozen objectives and purposes for the space program, ranging from advancing the sum total of human knowledge through creating and disseminating space and aeronautical technologies, mitigating the hazards posed by near-Earth objects and bioengineering, to improving ground propulsion technologies and developing technologies for improved cooling and heating. Apparently, we have a space program to improve the morning commute, or at least it is a result we should expect from the expenditure of billions of dollars and some of the nation's best technical talent. (Indeed, the advent of space-based geo-location and timing services arguably does just that, even if they were not created for those purposes.)

Unfortunately, creating a list of objectives and purposes is not the same as prioritizing them. Therein may lay the real problem. The space program does not suffer from a lack of vision, per se, but the lack of an agreed set of filters through which one can prioritize resource allocation. Reading the NASA Act, one cannot say for certain whether "ground propulsion systems research and development" is more important than "preservation of the role of the United States as a leader in aeronautical and space science and technology and in the application thereof to the conduct of peaceful activities within and outside the atmosphere." In fact, the structure of Section 20102 suggests that the former is more important than the latter, as the relevant paragraph states NASA's involvement in the former is required by the "general welfare," whereas international leadership is merely one of nine objectives which NASA's space and aeronautical activities shall be directed toward. Rather than pleading for a new vision, it might be preferable to argue for a list of priorities meaningful enough to provide useful guidance to NASA. The focus presumably created by a unifying vision is meaningless if it does not also guide resource allocation.

Introductory political science courses generally describe rationality as the ability to rank order preferences: A is more desirable than B; B is more desirable than C; and, so A must also be more desirable than C. To the degree that the NASA Act is unable to say over time that A is more important than B, which is more important than C, it fails to meet this elementary test of rationality.[75] It is a testament to the talent and dedication of the personnel working in the space program, whether as government employees, contractors, principal investigators, or educators that it has accomplished so much in the face of such problems.

The space program relies on a variety of mechanisms in an attempt to set priorities and impose a degree of reason on the program. Some work better than others. Within the space sciences, for example, the scientific community participates in a process of priority setting through the National Research Council's decadal surveys, which attempt to identify the most and least important things for the research community in the coming years. Other areas of scientific inquiry have adopted similar models. These work well within the enterprises that they apply to; they are not intended and do not function agency-wide or for the engineering disciplines that enable broader space activities. For that, the space program turns to its own strategic planning practices, which the National Research Council described as inadequate (through no fault of NASA's), and the political process by which democracies make collective decisions.

---

75 Research into preference formation and maintenance, particularly in groups, reveals that things are not so simple; people routinely prefer A to B, B to C, and C to A depending on the conditions in which choices are made. But, the point is still valid when it comes to establishing a single filter through which to set priorities.

Priority setting is no small task. It often involves winners and losers when it comes to allocating government resources, whether those resources are government contracts, subsidies, or simply the time and attention needed to establish or promote a supportive business environment. A "vision" for the space program that differs from the one we have now, general welfare promotion, may be helpful to this process if it requires one to set priorities.

The Apollo program did this. Reaching a definite location by a date certain, which was Kennedy's formulation, generally required NASA to set priorities, in the process ruling out some technical or programmatic options. One can think of program management in terms of three large variables: cost, schedule, and requirements. By prioritizing one or two variables, in this case schedule and modest but reasonably certain requirements (transport a person there and back) such a vision narrowed the trades that program managers could make. Fortunately, resources were forthcoming to deal with the consequences. Planetary exploration missions with definitive launch windows perform similar functions on a smaller scale today. The Pluto flyby, launched in 2006, followed a similar approach necessitated by the unique characteristics of Pluto's orbit.

This may be part of the reason that some often envision establishing a unifying vision a la Kennedy, with set destinations to be reached within a set time. Destination-driven missions create a forcing function in which program managers must set priorities if they expect to achieve their goals. Managers, scientists, and engineers must make decisions to meet deadlines rather than spend time exploring options. Indeed, that is the formulation that President Bush set for the VSE, when he directed NASA to develop a replacement for the Space Shuttle and demonstrate capability by 2014, return to the Moon by 2015, but no later than 2020, and lay the groundwork for sending people to Mars.[76]

Unfortunately, this kind of vision has several flaws. By setting deadlines, it automatically limits the trades that program managers may make. When the schedule is set, engineers may only compromise budgets or requirements in order to meet the goals. Moreover, its very strengths become weaknesses once the goal is achieved. Program managers naturally set priorities based on achieving the mission goal. As they discard less relevant technologies and capabilities, perhaps because they represent increased cost, additional schedule risk, or are not strictly necessary for achieving the immediate goal, they may also limit the value of the overall mission to society over the long term. The most visible benefit of such a program, the metric by which it will be measured, is actually accomplishing the mission, not the secondary or tertiary consequences that may have more social value over the long term, such as conducting new experiments or proving that technologies with wider applications actually work in space. Apollo, for example, resulted in incredible technology optimized for a single purpose; it did not leave behind the infrastructure one might associate with, or need for, moving human civilization into the solar system. In short, destination/schedule-driven visions are not conducive to a program that achieves multiple goals. Once the initial milestones are reached, there will be a natural inclination among policymakers to discontinue the activity due to its likely costs.

More recently, the Obama Administration proposed moving away from the destination-driven vision represented by the VSE. It cancelled the flagship technology programs associated with the destination-driven vision, intending instead to focus on operations in Low Earth Orbit and

---

76 National Aeronautics and Space Administration, *The Vision for Space Exploration,* February 2004, accessed at: http://www.nasa.gov/pdf/55583main_vision_space_exploration2.pdf

shift resources to new technology and capabilities. It hoped that those new technologies and capabilities would enable new, as yet unforeseen, activities in space, ultimately serving the national interest through the development of new industries and economic activity.[77] Members of Congress objected for several reasons. First, they rightly noted that the administration's proposal involved spending significant resources without a clear demonstration of purpose. Second, they were concerned that the administration's cancellation of the destination-driven Moon-Mars framework would return NASA to the aimlessness that allegedly dominated its decision-making prior to the loss of Columbia. Without the forcing function of a destination-driven vision, some expressed concern that the administration's approach would fail to set priorities.[78] As a result, Congress passed — and the President signed into law — annual authorization legislation resurrecting the keystone technology programs envisioned in the VSE, namely a heavy lift rocket and crew capsule, but not the vision that warranted their creation.

Still attempting to address the charge of aimlessness, NASA officials alternatively offered Mars, Martian moons, Lagrange points, and asteroids as destinations for NASA's efforts. By 2013, planning largely revolve around an asteroid mission, arguably in service of eventually sending people to Mars. Whereas the President in 2010 had called on NASA to visit one, NASA's fiscal year 2013 budget request called for NASA to bring an asteroid closer to Earth while sending astronauts to an asteroid in 2025.[79] In other words, NASA returned to a destination-driven vision. Even so, confusion about NASA's purpose persists.

---

77 See Eric R. Sterner, *Worthy of a Great Nation? NASA's Change of Strategic Direction* (Washington, DC: George C. Marshall Institute, April 2010).
78 Eric Sterner, "Congress, the White House, and Consensus: A Giant Leap Too Far," *Space Quarterly*, December 2011.
79 President Barack Obama, Remarks by the President on Space Exploration in the 21st Century, Merritt Island, Florida, April 15, 2010. See NASA's budget documents for fiscal year 2014, accessed at http://www.nasa.gov/news/budget/index.html.

At a recent hearing of the House Science Committee's Subcommittee on Space and Aeronautics, the Chairmen of the full Committee and Subcommittee both expressed skepticism about the asteroid mission, suggesting that it was unserious and could be a distraction from the ultimate goal of sending people to Mars.[80] The full committee's Ranking Minority Member, Eddie Bernice Johnson, noted in her prepared statement, "...despite clear policy direction in successive NASA Authorization Acts, NASA's human exploration program still has an air of tentativeness about it…if our nation's exploration program is to succeed, we need to have a clear roadmap to follow."[81] One of the witnesses, Paul Spudis, characterized the congressional reaction to the asteroid mission as "puzzlement" after the hearing.[82] It is no wonder that the calls for a unifying vision have not abated.

In the last decade, NASA has been guided by three different visions. First came the destination-driven Vision for Space Exploration, to be replaced by a technology focused approach signified by the Obama administration's 2010 budget request, which was modified again before settling on a destination-driven vision ostensibly focused on asteroids, presumably in the service of some larger space exploration framework. At that rate, the agency will be on to a fourth vision before the President's second term is finished. It is no way to run a railroad, much less a multi-billion dollar agency whose programs take years to bring to fruition, or a civil space program that was once an element of America's superpower status.

---

80 See the opening statements of Chairman Lamar Smith and Space Subcommittee Chairman Steven Palazzo, House Science Committee, Subcommittee on Space, Hearing "Next Steps in Human Exploration to Mars and Beyond," May 21, 2013. Available at: http://science.house.gov/hearing/subcommittee-space-next-steps-human-exploration-mars-and-beyond.
81 Hon. Eddie Bernice Johnson, "Opening Statement," before the House Committee on Science, Space, and Technology, Subcommittee on Space, Hearing "Next Steps in Human Exploration to Mars and Beyond," May 21, 2013. Available at: http://democrats.science.house.gov/sites/democrats.science.house.gov/files/documents/EBJ%20OPENING%20STATEMENTfinal%5BPRESS%5D.pdf.
82 Paul Spudis, "'Where, Why and How?' – Concerns of the Subcommittee on Space," Spudis Lunar Resources Blog, May 26, 2013, accessed at: http://www.spudislunarresources.com/blog/where-why-and-how-concerns-of-the-house-subcommittee-on-space/.

Uncertainty about the administration's asteroid proposal suggests that the current vision, however one chooses to characterize it, is not a consensus view that will guide the future of the space program. It may be the result of some commission, a tool for which policymakers often reach when the political process gets stuck. It may wait the outcome of a few more elections, which might bring about a different balance of political interests. Any process established to build consensus should consider alternatives for the future, each of which will involve certain tradeoffs among these different futures.

How then would we stack up competing visions? Vedda, Miller, and Adkins all designed capabilities-driven space programs based on three different unifying visions: settling the solar system, promoting commercial space development, and developing technologies for space-based applications (be they scientific, commercial, military, or geopolitical.) Clearly, that approach has much to commend it, as laid out in all three essays. Most appealing, perhaps, is the possibility that such an approach has the greatest potential to create success across a range of space activities: civil, scientific, commercial, and military. It would be a departure from calls for a destination-based vision in which policymakers define the "where" and "when" while the civil space program is expected to develop the "how" and employ it. In other words, a capabilities-based approach differs markedly from a destination-oriented vision for the agency. The former is more open-ended and may be less affected by the political and fiscal fluctuations that have dominated the agency's agenda since the end of the Apollo program. A destination-driven approach will get you from point A to point B. A capabilities-based approach will build you a car you can take anywhere and seek to develop the policies and programs that will also create the roads and gas stations you need to go there. That's great, so long as you need or want to travel.

That said, Adkins raises a potential challenge for capabilities-based approaches. As he points out in his discussion of a technology-focused program, one runs the risk that such efforts become ends in and of themselves, disconnected from the needs of the actors in the environment around them. Vedda and Miller, for example, both envision the creation of new industries and economic activities in, and derived from, space. They both set priorities to achieve these goals by demonstrating specific capabilities in consecutive order. Many of those potential industries will depend on the private sector for their creation and development. While beyond the scope of their essays, the question arises: what happens if the private sector finds no significant value in the proposed activities or that recovering resources in space is not economically viable? NASA was involved in man portable jet packs nearly a half a century ago: we still don't use them to commute to the office. They're not even an extreme sport. Antarctica may be a more useful example. Its territory contains significant resources, but human exploration has largely been limited to a handful of research activities and stations, even though the environment is arguably easier to use than space. Treaties made its economic development politically problematic while the cost and availability of resources elsewhere made such development economically pointless. To beat an analogy to death, a capabilities-based approach to build superhighways and truck stops in Antarctica would be a waste; there's nobody there to use them, nor are they likely to get to Antarctica in meaningful numbers just because the technology to do so exists. (It should be noted that there are tourists in Antarctica, and soon likely will be in space).

A government-led, capabilities-focused program is essentially an attempt to resolve this chicken-and-egg problem. As Vedda and Miller envision, by building the infrastructure, government makes it easier for the private sector to step in. But what happens if the government gets it wrong? What happens if the government "builds it" and "nobody comes."

Most decisions involve opportunity costs, i.e., the cost of foregoing course A in order to pursue course B, or at least the marginal difference in benefit between the two. Merriam-Webster's on-line dictionary explains opportunity cost as "the added cost of using resources (as for production or speculative investment) that is the difference between the actual value resulting from such use and that of an alternative (as another use of the same resources or an investment of equal risk but greater return)."[83] Hopefully, the benefits of choosing course B are preferable to those of option A, which is why it's chosen. It may be easier to ask yourself what you had to forego in order to choose one option, rather than the other. The space program is experiencing this dilemma currently, as policymakers wrestle with tradeoffs between the Space Launch System and financial support for privately-owned and operated human spaceflight capabilities. For that matter, money spent on human spaceflight is not available for space or earth science. Nor is it available for biomedical research at the National Institutes of Health. Because budgeteers face such tradeoffs, the political system does not reward the development of extensive capabilities without a purpose already in mind. Bluntly, the government has not been willing to spend limited tax resources on a "build it and they will come" approach.

Pace does not address the issue, but his approach to maximizing America's geopolitical return from its space program highlights the point. Maximizing the soft power benefits from our space program requires greater attention to the needs, interests, and desires of current and potential allies and friendly nations in space. In all likelihood, those countries

---

83 Meriam-Webster.com, "opportunity cost," http://www.merriam-webster.com/dictionary/opportunity%20cost.

will seek technological and economic benefits, in addition to the scientific and geopolitical benefits of cooperating with the United States. The United States might have to forego some of those technological and economic benefits in order to accommodate potential partners and reap the soft power benefits of cooperation. It might also have to accept a modicum of risk that increased technological capabilities resulting from such cooperation could be turned to military purposes, as most space technology—and certainly experience in space research, development, testing, engineering, and operations—could have military benefits.

Conversely, a capabilities-based approach that focuses on development or commercialization must subordinate other goals, such as geopolitics. It would not do to invest billions of tax dollars helping finance the development of private, U.S. space capabilities if, at the same time, the government was pursuing cooperative projects that encouraged other nations to develop similar capabilities in the pursuit of soft power, thus reducing the potential customer base, and creating new competitors, for U.S. service providers. Thus, the programs that a geopolitical approach favors become less important, meaning the United States may have to forego the important strategic benefits that they represent in order to concentrate its efforts on creating capabilities for other purposes.

Fortunately, tradeoffs are not always so zero-sum. To be sure, devising a program with one vision in mind will contribute to the achievement of other broad goals. As Miller points out, Mahan laid this out for naval power in the 19th century. An advanced and capable navy required strong commercial maritime foundations.[84] Meanwhile, a strong navy helped protect the interests of a commercial fleet. Similar relationships are likely

---

84 The United States, for example, has been experimenting for the last forty years with the ability to maintain a strong navy despite the absence of strong, commercial, maritime foundations. The U.S. shipbuilding industry has withered since World War II—most ships are built overseas, sail under foreign flags, and have multinational crews.

135

to hold true for space. For example, government funding for the development of space launch vehicles and payloads led to exploration of the radiation belts surrounding Earth. James Van Allen, for whom they are named, did early technical work on the U.S. program to exploit the V-2 rocket and the Explorer I, which carried his Geiger counter into orbit, was launched aboard a Jupiter-C rocket developed by the Army Ballistic Missile Agency. Similarly, the technological and professional skills developed for the Apollo program — initiated for geopolitical purposes — also contributed to the foundations for America's scientific exploits in space. These relationships are immensely complex and well beyond the discussions contained here, but they are very real.

The fact that taking steps in advancing one goal does not always preclude progress against another goal does not mean that a hierarchy of preferences is unnecessary. Those opportunity costs are still real. Establishing a focus would nominally create a filter through which we process decisions; in effect it would help policymakers establish a rank ordering of preferences. When two goals conflict, some principle would guide choices. Policymakers, program managers, budgeters, and the like would have some means of making informed decisions. A capability-driven vision may accomplish this by following a technology roadmap. But that, in and of itself, does not answer the question: what happens if the roadmap proves to be a roadmap to nowhere.

Unsurprisingly, none of our authors proposed breaking NASA up and parceling the pieces out to other departments or agencies. Instead, our authors proposed changing the agency, an implicit acknowledgment that NASA as currently constituted is not ideally suited to achieve the visions they explored for it. Indeed, the proposition seems counterintuitive and we did not ask authors explicitly to consider it. But, in the interests of taking the thought experiment a step further, let's consider it.

Assume for a moment that a consensus grows around the existing set of programs, i.e., that the status quo, level-of-effort vision persists. Our civil space program exists to achieve the myriad and weakly connected goals laid out in the NASA Act. The NASA Act states that a single civilian agency shall carry out the activities required by the general welfare. At the time of NASA's creation, the repository of talent capable of engaging in space activities was limited; there was a certain logic in bringing it together. Today, the situation is different. The engineering and scientific capabilities needed to engage in space activities are widely spread, both in the government and the private sector.

Consider the first objective for NASA as laid out in the "general welfare" objective that warrants its existence. The agency is supposed to expand human knowledge of the Earth and phenomena in the atmosphere and space. It has done this admirably and with remarkable success using what had been its unique talents with space-based systems. But it is no longer the only government agency studying the planet's global environment or outer space. The U.S. Air Force and National Science Foundation conduct or fund research into space phenomena, albeit for different reasons. The U.S. Navy, National Oceanic and Atmospheric Administration, National Weather Service, and U.S. Coast Guard all collect data and study phenomena in the oceans and atmosphere. The U.S. Geological Survey, Environmental Protection Agency, National Geospatial Intelligence Agency, and Department of Agriculture all collect and use data for similar purposes. Few of them design, build, manage, or operate space systems, but with a transfer of resources and talent from NASA, there should be no doubt that they could as needed. Integrating metadata, which many of those agencies already collect, with that available from space systems will likely improve the quality of research in the future. It may be time to treat space as a medium and means to an end, rather than an end in itself that requires a separate agency. With modest changes to the missions of those agencies and the divvying up of NASA resources among them, those agencies could conceivably add to the sum total of knowledge as well as, if not more efficiently than, a single agency charged with the mission simply because circumstances during the 1950s warranted it. We do not have departments or agencies responsible for the land, sea or air. Does it still make sense to have one for this particular aspect of space activity?

Things may be different with respect to NASA's responsibilities to improve the usefulness, performance, speed, safety, and efficiency of aeronautical and space activities. The Department of Defense already does this, but since World War II its research and development activities in these areas have become highly specialized for military missions. No other agency performs this kind of work today. A single agency may be best suited to it for unspecialized purposes, much as NACA focused on across-the-board improvements in aeronautics.

One of the most challenging, and amorphous, objectives for the space program as currently constituted is "the preservation of the role of the United States as a leader in aeronautical and space science and technology and in the application thereof to the conduct of peaceful activities within and outside the atmosphere." At the same time, the Act states international cooperation serves the general welfare, without specifying how.

As unifying visions go, this is not very helpful. The NASA Act does not define leadership and policymakers are generally forced to substitute their personal notions of international leadership and the metrics that might go with it. These may include spending levels, high-profile achievements (such as sending people and payloads to distant targets), and basic capabilities (such as launching people to orbit). If those metrics are how we measure leadership, there is no requirement that they reside in a single agency. We already consider aggregate spending across agencies when we compare the space programs of foreign countries and capabilities can be spread across a range of departments and agencies, each of which has a primary task other than maintaining (or restoring) U.S. leadership in space. Moreover, the geopolitical agenda that Pace lays out, which is well suited to U.S. leadership, would be difficult to coordinate through a multiplicity of departments and agencies. Quite simply, other countries might be reluctant to work with U.S. military services or the relatively mid- or low-level officials that would have responsibilities for space programs in larger departments. In the same vein, cooperation can involve a host of competing issues: ranging from relative capabilities of potential partners, the reliability/volatility of political relationships, technical and cultural differences, proliferation concerns, technology transfer, and human rights practices. A single agency might be best suited to nurturing the cooperative relationships, lest agencies with narrower mission-oriented focuses fail to consider that range of issues.

Just this cursory review suggests that if we started from scratch and were given the current "general welfare" vision, we would probably create a civil space agency. It might not look like the NASA of today. Science activities might be spread across a range of departments and agencies, which would see their capabilities to use space enhanced. But, the general goals of developing new technologies and vehicles and pursuing international leadership would likely reside in a single agency. That said, it might look more like a cross between NACA, which was largely an engineering entity, and DARPA, which has been unique in its own right. (Of course, DARPA's historic successes prompted the government to try and replicate it across several departments.) It would not have to be an agency that operated space systems, such as the Space Shuttle or Space Station, over long periods of time. Instead, it might rely on other entities to conduct such operations.

The point of this little excursion is not to defend, or attack, NASA. The discussion is hardly rigorous or sophisticated enough to be used in that way. Instead, I want to suggest that the makeup and organization of whatever space program we have must change in order to best suit whatever vision selected for it. Too often, exercises in charting a future for NASA work the other way: the vision is offered as something that NASA needs. Rather than having the tail wag the dog with a broad debate about what best will serve the agency, discussion about future visions should begin with questions about the national interest. What best serves the country? An integrated strategy to settle the solar system? Using space to serve geopolitical interests? A focus on technology development? A narrow focus on reducing prices of foundational capabilities, such as access? Promoting aerospace activities simply because they serve the general welfare?

Once policymakers answer those questions, they can begin a second debate over how best to achieve them. In conducting their thought experiments about how to achieve the visions we gave them — and without reference to the capabilities, institutions, or history that we currently have — our authors designed something other than the agency we have today. In short, today's NASA may not be best suited to U.S. needs in the 21st century. Significant, and difficult, changes may be in order.

This does not mean that any of our authors would recommend the changes we asked them to explore. The truth is that policy analysts and policymakers do not start with a clean sheet of paper when they consider options. They must deal with reality, as it exists, with the competing interests, talent bases, capabilities, legacies, and expectations that the United States inherited from all that has gone before, and not created for the future. Our authors are aware of that and might well make different recommendations to policymakers when it comes to improving the space program we have. But, their contributions highlight a critical aspect of the process that is too often lacking. As our space program moves forward, it's critical that we not simply start with where we are and look to make things better, but that we have a clear eye on where we want to be.

About the Authors

William B. Adkins is president of Adkins Strategies, LLC, a Washington, DC-based space and defense consulting firm. He has more than 27 years' experience in the national security and civil space arenas, including extensive experience in both the Executive Branch and Congress. Prior to forming Adkins Strategies, Bill served on the professional staff of the House Science Committee from 2000-2006. In 2001, he was appointed as the staff director of the Space and Aeronautics Subcommittee where he led the subcommittee's legislative and oversight activities of NASA and other civilian space activities. Before joining the staff of the U.S. House of Representatives, he was a Legislative Assistant and National Security Fellow in the U.S. Senate where he handled national security issues. Prior to Capitol Hill, he worked at the National Reconnaissance Office (NRO) where he served as a project manager in the Advanced Systems and Technology Directorate. From 1990-1993, he served as an engineer in a System Program Office at the NRO. At the Naval Research Laboratory from 1986-1990, he participated in the design, development and operations of several defense and scientific space missions. Born in Washington, D.C., Bill earned a Bachelor of Science degree in Mechanical Engineering from George Washington University and has completed graduate courses in Electrical Engineering at George Washington University. He has also studied at Harvard University and Johns Hopkins University.

Charles M. Miller is the President of NexGen Space LLC, which provides client-based services at the intersection of commercial, civil and national security space and public policy. Miller's clients have included the U.S. Air Force, DARPA, NASA, the White House Office of Science & Technology Policy, as well as many private firms. He served as NASA Senior Advisor for Commercial Space from 2009 to 2012. Miller is also the co-founder of Nanoracks LLC, the former President and CEO of Constellation Services International, Inc., and the founder of ProSpace, which was called "The Citizens' Space Lobby."

Scott D. Pace is the Director of the Space Policy Institute and a Professor of Practice in International Affairs at George Washington University's Elliott School of International Affairs. His research interests include civil, commercial, and national security space policy. From 2005-2008, he served as the Associate Administrator for Program Analysis and Evaluation at NASA. Prior to NASA, Dr. Pace was the Assistant Director for Space and Aeronautics in the White House Office of Science and Technology Policy (OSTP).

Eric R. Sterner is a Fellow at the George C. Marshall Institute and teaches in Missouri State University's Graduate Department of Defense and Strategic Studies. In addition to experience in the private sector, he held senior staff positions at the Committee on Science and the Committee on Armed Services in the U.S. House of Representatives and served in the Defense Department and as NASA's Associate Deputy Administrator for Policy and Planning. His work has appeared in a number of publications, including *The Washington Post, Aviation Week and Space Technology, Strategic Studies Quarterly, Comparative Strategy, The Washington Quarterly*, and *Space News*.

James A. Vedda is a senior policy analyst at the Aerospace Corporation's Center for Space Policy & Strategy in Arlington, Virginia, where he has been performing research and analyses on national security, civil, and commercial space issues since 2004. Previously, he spent six and a half years assigned to the Office of the Secretary of Defense working on space policy and homeland defense issues. Before that, he was an associate professor in the Department of Space Studies at the University of North Dakota, where he taught courses on civil, commercial, and military space policy to undergraduate and graduate students. As one of the founding members of the department, he helped create the curriculum for the Master of Science in Space Studies degree. He holds a Ph.D. in political science from the University of Florida and a master's degree in Science, Technology, and Public Policy from George Washington University. He is the author of *Choice, Not Fate: Shaping a Sustainable Future in the Space Age* (December 2009) and *Becoming Spacefarers: Rescuing America's Space Program* (June 2012). His writing also has appeared in book chapters and in journals such as *Space Policy*, *Space News*, *Astropolitics*, *Space Times*, *Ad Astra*, and *Quest*. He has presented conference papers for the International Astronautical Federation, the American Institute of Aeronautics & Astronautics, the Midwest Political Science Association, the NASA History Office, and the National Air & Space Museum, and commentary for CNN, *The Space Show*, and others.

About the George C. Marshall Institute

The George C. Marshall Institute (www.marshall.org) was established in 1984 as a nonprofit 501(c)(3) corporation to conduct technical assessments of scientific issues with an impact on public policy.

In every area of public policy, from national defense, to the environment, to the economy, decisions are shaped by developments in and arguments about science and technology. The need for accurate and impartial technical assessments has never been greater. However, even purely scientific appraisals are often politicized and misused by interest groups.

The Marshall Institute seeks to counter this trend by providing policymakers with rigorous, clearly written and unbiased technical analyses on a range of public policy issues. Through briefings to the press, publication programs, speaking tours and public forums, the Institute seeks to preserve the integrity of science and promote scientific literacy.

We publish reports, host roundtables, workshops and collaborate with institutions that share our interest in basing public policy on scientific facts.